Structural Engineering Institute
American Society of Civil Engineers

Guideline for Structural Condition Assessment of Existing Buildings

This document uses both Système International (SI) units and customary units.

Published by the American Society of Civil Engineers
1801 Alexander Bell Drive
Reston, Virginia 20191-4400

ABSTRACT
Changing economic conditions, concerns for historic preservation, emphasis on fully utilizing conveniently located structures, space shortages, and increasing cost of materials and products used in the construction of new buildings have resulted in a need to evaluate and more fully utilize the existing building inventory. To this end, the standard *Guideline for Structural Condition Assessment of Existing Buildings (ASCE 11-90)* was developed to provide the design community with guidelines for assessing the structural conditions of existing buildings constructed of combinations of material including concrete, masonry, metals, and wood. This edition (SEI/ASCE 11-99) replaces ASCE 11-90. It consists of an overview of preliminary and detailed assessment procedures, of materials properties and test methods, and of evaluation procedures for various physical conditions of the structure. The standard is not intended to be inclusive or prescriptive but is expected to serve as a resource document for engineers, owners, and regulatory officials.

Library of Congress Control Number: 00-132816

Photocopies. Authorization to photocopy material for internal or personal use under circumstances not falling within the fair use provisions of the Copyright Act is granted by ASCE to libraries and other users registered with the Copyright Clearance Center (CCC) Transactional Reporting Service, provided that the base fee of $8.00 per article plus $.50 per page is paid directly to CCC, 222 Rosewood Drive, Danvers, MA 01923. The identification for ASCE Books is 0-7844-0432-1/00/$8.00 + $.50 per page. Requests for special permission or bulk copying should be addressed to Permissions & Copyright Dept., ASCE.

Copyright © 2000 by the American Society of Civil Engineers,
All Rights Reserved.
Library of Congress Control No: 00-132816
ISBN 0-7844-0432-1
Manufactured in the United States of America.

STANDARDS

In April 1980, the Board of Direction approved ASCE Rules for Standards Committees to govern the writing and maintenance of standards developed by the Society. All such standards are developed by a consensus standards process managed by the Management Group F (MGF), Codes and Standards. The consensus process includes balloting by the balanced standards committee made up of Society members and nonmembers, balloting by the membership of ASCE as a whole, and balloting by the public. All standards are updated or reaffirmed by the same process at intervals not exceeding 5 years.

The following Standards have been issued.

ANSI/ASCE 1-82 N-725 Guideline for Design and Analysis of Nuclear Safety Related Earth Structures

ANSI/ASCE 2-91 Measurement of Oxygen Transfer in Clean Water

ANSI/ASCE 3-91 Standard for the Structural Design of Composite Slabs and ANSI/ASCE 9-91 Standard Practice for the Construction and Inspection of Composite Slabs

ASCE 4-98 Seismic Analysis of Safety-Related Nuclear Structures

Building Code Requirements for Masonry Structures (ACI 530-99/ASCE 5-99/TMS 402-99) and Specifications for Masonry Structures (ACI 530.1-99/ASCE 6-99/TMS 602-99)

ASCE 7-98 Minimum Design Loads for Buildings and Other Structures

ANSI/ASCE 8-90 Standard Specification for the Design of Cold-Formed Stainless Steel Structural Members

ANSI/ASCE 9-91 listed with ASCE 3-91

ASCE 10-97 Design of Latticed Steel Transmission Structures

SEI/ASCE 11-99 Guideline for Structural Condition Assessment of Existing Buildings

ANSI/ASCE 12-91 Guideline for the Design of Urban Subsurface Drainage

ASCE 13-93 Standard Guidelines for Installation of Urban Subsurface Drainage

ASCE 14-93 Standard Guidelines for Operation and Maintenance of Urban Subsurface Drainage

ASCE 15-98 Standard Practice for Direct Design of Buried Precast Concrete Pipe Using Standard Installations (SIDD)

ASCE 16-95 Standard for Load and Resistance Factor Design (LRFD) of Engineered Wood Construction

ASCE 17-96 Air-Supported Structures

ASCE 18-96 Standard Guidelines for In-Process Oxygen Transfer Testing

ASCE 19-96 Structural Applications of Steel Cables for Buildings

ASCE 20-96 Standard Guidelines for the Design and Installation of Pile Foundations

ASCE 21-96 Automated People Mover Standards—Part 1

ASCE 21-98 Automated People Mover Standards—Part 2

SEI/ASCE 23-97 Specification for Structural Steel Beams with Web Openings

SEI/ASCE 24-98 Flood Resistant Design and Construction

ASCE 25-97 Earthquake-Actuated Automatic Gas Shut-Off Devices

DEDICATION TO JEROME S. B. IFFLAND

This second edition of ASCE Standard Guideline for Structural Condition Assessment of Existing Buildings is dedicated to the memory of Jerome S. B. Iffland, who passed away in February 1995. Jerry was a founding member of the committee established by ASCE in 1983 and served as its Vice-Chairman until his death. While providing overall direction to the committee based on his extensive professional experience with assessment and rehabilitation of structures, he was an active worker preparing important sections of the standard ready for consensus review. The completion of the 1990 edition of the standard was due in large part to his efforts.

Jerry will be especially remembered for drafting parts of this second edition of ASCE 11 while he was very ill. Much of this work was done while he was in the hospital. A special thanks goes to Jerry's wife, Helen, who assisted him in working on the standard during this difficult period.

It is with great pleasure that the ASCE Standards Committee on Structural Condition Assessment of Existing Buildings dedicates this standard to the memory of Jerry Iffland. He was an inspiration to the committee, setting an example that guided our work even after his death.

FOREWORD

Changing economic conditions, concern for historic preservation, emphasis on fully utilizing conveniently located structures, space shortages, and increasing cost of materials and products used in the construction of new buildings have resulted in a need to evaluate and more fully utilize the existing building inventory. Particularly in older cities, emphasis has shifted from replacement to preservation, rehabilitation, and strengthening of existing buildings.

New processes resulting in changes of building systems and business equipment frequently impose greater loads on an existing building structure and may require additional openings and restructuring. More stringent building code provisions for design load requirements or improved seismic resistance may also demand retrofitting of structural reinforcement. Also, any known site conditions should be reviewed to determine if modifications to structural systems are required.

Adaptive reuse, rehabilitation, and improvement of existing buildings all require an accurate assessment of present building performance and capability for use by owners, designers, building officials, and contractors. Current data for assessment is dispersed and not readily available to many of those making technical decisions. Such information has been compiled and subjected to a consensus review and approved by this committee to provide the design community with a resource standard on building condition assessment for selected materials and for other areas related to the structural performance of buildings. To that end, this Standard Guideline for Structural Condition Assessment of Existing Buildings has been prepared for use by qualified professional engineers and regulatory officials. This edition replaces ASCE 11, first published in 1990.

This Standard is not intended to be inclusive or prescriptive. Methods and procedures are presented as a resource for reference purposes. Other methods and procedures are not only permissible, but are encouraged, so long as they are deemed reliable and sufficient comparisons are available with other recognized methods.

Inasmuch as interpretation of the results of the evaluation must be based on the professional experience and judgement of the practitioner, it is not a part of this Standard.

Utilization of this guideline may involve hazardous materials, operations and equipment. It does not purport to address all of the safety problems associated with its use. It is the responsibility of the user of this guideline to establish appropriate safety and health practices and determine the applicability of regulatory limitations prior to use.

The material presented in this publication has been prepared in accordance with recognized engineering principles. This Standard Guideline should not be used without first securing competent advice with respect to its suitability for any given application. The publication of the material contained herein is not intended as a representation or warranty on the part of the American Society of Civil Engineers, or of any other person named herein, that this information is suitable for any general or particular use or promises freedom from infringement of any patent or patents. Anyone making use of this information assumes all liability from such use.

ACKNOWLEDGMENTS

The Structural Engineering Institute (SEI) and the American Society of Civil Engineers (ASCE) gratefully acknowledge the work of the SEI standards committee Structural Condition Assessment and Rehabilitation of Buildings. This committee comprises individuals from many backgrounds including: consulting engineering, research, construction industry, education, government, design and private practice.

This revision of the standard began in 1995 and incorporates information as described in the commentary.

This Standard was prepared through the consensus standards process by balloting in compliance with procedures of ASCE's Codes and Standards Activities Committee. Those individuals who served on the Structural Condition Assessment and Rehabilitation of Buildings Standards Committee are:

Kenneth Adams
Appupillai B. Baskaran
Prodyot K. Basu
Carl A. Baumert, Jr.
Kasi V. Bendapudi
Donald A. Berg
Viggo Bonnesen
James Brown
W. G. Corley
David A. Deress
Richard M. Gensert
Stephen H. Getz
Satyendra K. Ghosh
V. Gopinath
Melvyn Green
James A. Hill
Charles J. Hookham
Nestor R. Iwankiw
Dennis Kutch
Michael J. Lavallee
Arnold N. Lowing
Charles R. Magadini
Rusk Masih
Robert R. McCluer
Paul L. Millman
Ronald J. Morony
George R. Mulholland
Antonio Nanni
William R. Nash
Joseph F. Neussendorfer
Dai H. Oh
Celina U. Penalba
James Pielert, Chair
Denis C. Pu
Janah A. Risha
William D. Rome
Robert J. Schaffhausen
Glenn R. Smith, Jr.
Brian E. Trimble
Donald R. Uzarski

C. R. Wagus
Frederick L. Walters
Ralph J. Warburton
Joseph A. Wintz, III

Subcommittee on Structural Condition Assessment of Existing Buildings

Charles E. Bacchus
Prodoyt K. Basu
Carl A. Baumert
Kasi V. Bendapudi
Viggo Bonnesen
James Brown
James R. Clifton (Deceased)
W. G. Corley
Robert H. Falk
Richard M. Gensert
Satyendra K. Ghosh
V. Gopinath
Melvyn Green, Secretary
Michael C. Henry
James A. Hill
Charles J. Hookham
Jiin Long Huang
Samuel L. Hunter
Jerome S.B. Iffland (Deceased)
Nestor R. Iwankiw
Wen-Chen Jau
Charles J. Kanapicki
Jack Kayser
Mahadeb Kundu
Dennis Kutch
Jim E. Lapping
Arnold N. Lowing
Charles R. Magadini
Richard McConnell
Kirk I. Mettam
George R. Mulholland

Joseph F. Neussendorfer
Dai H. Oh
Anthony J. Pagnotta
Celina U. Penalba
James Pielert, Chair
Laurence Pockras
Denis C. Pu
Janah A. Risha
Willam D. Rome
Robert J. Schaffhausen
Paul A. Seaburg
Glenn R. Smith, Jr.
Clayton R. Steele
Brian E. Trimble
Frederick L. Walters
Ralph J. Warburton
Thomas G. Williamson
Joseph A. Wintz, III
Wade W. Younie

Subcommittee on Condition Assessment of the Building Envelope

Appupillai B. Baskaran
Carl A. Baumert, Jr. Chair
Robert J. Beiner
Donald A. Berg
George G. Cole
David A. Deress
Robert H. Falk
Stephen H. Getz
V. Gopinath
Mark Graham
Melvyn Green
Michael C. Henry
James A. Hill
Samuel L. Hunter
Wen-Chen Jau
Mahadeb Kundu

Dennis Kutch
Roger A. LaBoube
Jim E. Lapping
Michael J. Lavallee
Rusk Masih
Kirk I. Mettam
Paul L. Millman
Ronald J. Morony
George R. Mulholland
Joseph F. Neussendorfer
Anthony J. Pagnotta
James Pielert
Denis C. Pu
Janah A. Risha
William D. Rome
Robert J. Schaffhausen
Paul A. Seaburg
Clayton R. Steele
Brian E. Trimble
Donald R. Uzarski
C. R. Wagus
Richard Walker
Frederick L. Walters, Secretary
Joseph A. Wintz, III
Wade W. Younie

CONTENTS

		Page
STANDARDS		iii
DEDICATION		v
FOREWORD		vii
ACKNOWLEDGMENTS		ix

1.0	General	1
	1.1 Scope and Intent of Standard	1
	1.2 Purpose of Assessment	1
	1.3 Qualifications and Equipment	1
	1.4 Agreements	1
	1.5 Definitions	2
	1.6 References	2
2.0	Assessment Procedure	3
	2.1 General	3
	2.2 Approach	3
	2.3 Preliminary Assessment	3
	2.4 Detailed Assessment	6
3.0	Structural Materials Assessment	8
	3.1 Purpose and Scope	8
	3.2 Condition Assessment of Structural Concrete	8
	3.3 Condition Assessment of Metals	16
	3.4 Condition Assessment of Masonry	20
	3.5 Condition Assessment of Wood	27
	3.6 References	33
	3.7 Tabulation of Test Methods	35
	3.8 Supplemental References	89
4.0	Evaluation Procedures and Evaluation of Structural Materials and Systems	93
	4.1 Evaluation Procedures	93
	4.2 Evaluation of Structural Concrete	94
	4.3 Evaluation of Metals	104
	4.4 Evaluation of Masonry	108
	4.5 Evaluation of Structural Wood	115
	4.6 Other Materials	126
	4.7 Evaluation of Components and Systems	126
	4.8 Interpretation	133
	4.9 References for Evaluation Procedures	133
5.0	Report on Structural Condition Assessment	137
Appendix A—Report of Structural Condition Assessment		140
Appendix B—Organization References		141
Index		143

Guideline for Structural Condition Assessment of Existing Buildings

1.0 GENERAL

1.1 SCOPE AND INTENT OF THE STANDARD

The intent of this Standard is to provide guidelines and methodology for assessing the structural condition of existing buildings constructed of combinations of materials including concrete, metals, masonry, and wood. This Standard establishes the assessment procedure including investigation, testing methods, and format for the report of the condition assessment.

Since any evaluation will involve engineering judgment and contains factors that cannot be readily defined and standardized, a Section providing guidance for evaluations is also included. This Section must be used by the professional engineer as part of an engineering evaluation. The Standard is intended as a guide to the engineer in providing comprehensive information for clients such as building owners, prospective purchasers, tenants, regulatory officials, and others.

Dimensions and quantities in this Standard are expressed in lb. units followed by conversion to SI units in parentheses.

1.2 PURPOSE OF ASSESSMENT

Structural condition assessment of an existing building can be undertaken for a number of purposes. These can include developing a performance report, establishing building use, serviceability, code compliance, life safety, durability, historic preservation, or a number of special purposes based on the specific building and its current or proposed occupancy or function. The engineer should consider the possibility of the presence of hazardous material when assessing an existing building, advise the client as necessary, and take or recommend appropriate precautions.

1.3 QUALIFICATIONS AND EQUIPMENT

1.3.1 Personnel Qualifications

All personnel involved in the assessment shall possess the technical qualifications, including practical experience, education, and professional judgment required to perform the individual technical tasks assigned. Interpretation of results and conclusions shall be performed by a registered professional engineer qualified in the appropriate discipline.

1.3.2 Equipment

Equipment shall be obtained as appropriate to accomplish or perform the various tests and inspection methods specified in the Standard. All equipment shall be in good working order. For equipment that can be calibrated, reports of calibration shall be available.

1.4 AGREEMENTS

1.4.1 Services

The scope of services for the structural condition assessment shall be defined by the qualified professional engineer, and all conditions, applicable codes and standards, and services shall be mutually agreed upon by the client and professional engineer. Services may include one or more of the identified purposes of the assessment and may involve more than one specific building component or system covered in the subsections of this Standard. Evaluation and acceptance criteria should be defined as part of the agreement. Clients should be advised of the possibility of hazardous material being present in an existing building unless a previous investigation has been completed.

1.4.2 Compensation

The client and the qualified professional engineer shall mutually agree upon the engineer's compensation for the services specified in Section 1.4.1. The agreement shall specify if it includes reimbursable expenses such as testing laboratory costs. Compensation provisions shall consider that, after review of the preliminary data, additional investigations may be required, thereby changing the scope of the project.

1.4.3 Authority and Accessibility

The agreement shall clearly identify the scope and authority of the professional engineer to perform the necessary investigation and tests and shall assure access to the site, the building, the various portions of the building requiring investigation, and drawings and documents required to perform the condition assessment. Responsibility for removal and repairs to finishes and building elements that may result from the evaluation, test or inspection procedures should be identified.

1.4.4 Liability

The extent of liability, if any, expressly accepted by the professional engineer shall be stated in the agreement. Requirements for liability coverage shall be clearly documented.

1.5 DEFINITIONS

The following definitions are provided to assure uniform understanding of some selected terms as they are used in this Standard:

Assessment: Systematic collection and analysis of data, evaluation, and recommendations regarding the portions of an existing building which would be affected by its proposed use.

Attributes: Building performance features or characteristics such as fire safety or behavior under seismic loading.

Building: An enclosed or partially enclosed structure for occupancy by personnel, material or equipment.

Client: Individual or organization having responsibility or control over the building being assessed.

Component (Nonstructural): A component whose original primary function was other than to support vertical or lateral loads that may be imposed by the building except for its mass. These may include ceilings, stairs, parapets, ornamentation, cornices, overhangs, equipment, flagpoles, chimneys, in-fill walls, and handrails.

Component (Structural): A portion of a building that, for the purposes of evaluation, can be isolated from the remainder of the building and which possesses a capability to support, resist, or transfer loads. These components include, for example, roof decks, floors, walls, elevator cores, stair shafts, frames, foundation structures, piles and piers.

Computational Analysis: A quantitative evaluation of a building component or system by or under the direction of a professional engineer.

Connection: The fastening between components or other parts of the building which either by design or as a result of construction are capable of transferring loads or forces. A "connector" is a connecting device.

Destructive Testing: Process of observing, inspecting, and/or measuring the properties of materials, components, or system in a manner which may change, damage, or destroy the properties or affect the service life of the test specimen. Destructive testing of portions of a building may be carried out on the building in situ or on samples removed and tested in the laboratory.

Evaluation: The process of determining the structural adequacy or the building or component for its intended use and/or performance. Evaluation by its nature implies the use of personal and subjective judgment by those functioning in the capacity of experts.

Inspection: The activity of examining, measuring, testing, gauging, and using other procedures to ascertain the quality or state, detect errors, defects, or deterioration and otherwise appraise materials, components, systems, or environments.

Nondestructive Testing: Assessment of a material, component, or system without altering properties or impairing future service.

On-Site Load Testing: Testing carried out on building systems on-site to determine structural performance characteristics.

Professional Engineer (Engineer): A person duly licensed and registered to practice engineering in the governmental jurisdiction.

System: An assemblage of structural components within a building that, by contiguous interconnection form a path by which external and internal forces applied to the building, are delivered to the ground.

Testing Laboratory: A laboratory in which the characteristics or performance of materials and products are measured, examined, tested, calibrated, or otherwise determined.

1.6 REFERENCES

References are significant to the use of this Standard. The more the user makes use of these references and becomes familiar with them, the more valuable the Standard will be. It should be noted that many references are taken from the field of bridge engineering even though the subject of this Standard is buildings. Much of the research on material problems, tests on samples of the materials, and nondestructive testing has been in the bridge engineering discipline. Bridge structural systems are exposed and therefore subject to more severe weathering, have less redundancy, and typically cannot be abandoned as they deteriorate. These references are valuable since the same materials are used in both bridges and buildings.

2.0 ASSESSMENT PROCEDURE

2.1 GENERAL

The process of assessing the structural condition of an existing building consists of assembling and systematically analyzing information and data regarding the building or portions thereof in order to determine the structural adequacy. A general procedure for conducting a structural assessment of existing buildings is shown in Fig. 2-1 and described herein. The procedure is not necessarily complete and the ordering of the steps in the procedure depends on the specific purpose of the investigation, the availability of design and construction documents, and the number of buildings encompassed in the investigation. The precise procedure for the condition assessment of each individual building shall be determined by the professional engineer in charge of the investigation. This Standard guideline does not preclude the use of other procedures.

2.2 APPROACH

Due to the potential cost of a comprehensive structural assessment for an existing building, a multilevel approach is generally recommended. The basic process entails a preliminary assessment followed by a detailed assessment, if required. When it has already been determined that the building requires a detailed assessment, the preliminary assessment may be skipped. If the scope of the investigation encompasses a large inventory of buildings, screening criteria should be established and the inventory screened to select buildings that will be included in the preliminary assessment. Potential screening factors such as date of construction, building area, acquisition cost, importance, and occupancy can be used to reduce and prioritize the inventory for preliminary assessment. The purpose of inventory reduction is to minimize or eliminate unnecessary detailed investigations and to keep the scope of work within reasonable limits.

2.3 PRELIMINARY ASSESSMENT

The preliminary assessment provides the initial analytical data for estimating the structural adequacy of an existing building and for establishing the need and priority for a more detailed analysis. The basic steps include: (1) available documents review; (2) site inspection; (3) preliminary analysis; and (4) preliminary findings and recommendations.

2.3.1 Available Documents Review

For the preliminary assessment, design information such as drawings, design criteria, soil investigations, calculations, and other information will, if available, permit an easier and more accurate evaluation and expedite the site inspection. When possible, such information should be assembled prior to the inspection. When such information is not available, the engineer must obtain the required information and data at the site.

2.3.2 Site Inspection

A site inspection enables the engineer to confirm the correctness of the existing design information and to assess the structural condition of the building. Any evidence of structural modification, deterioration of materials, discrepancies in documentation, weakness in structural members or connections, settlement or foundation problems, or unusual structural features should be noted. The inspection team should include at least one structural engineer experienced in structural evaluation. Also, personnel familiar with the building, such as the building supervisor, should accompany the inspection team to provide access to various areas within the building, describe functional requirements, and point out any known areas of modification, deterioration, and damage.

2.3.3 Preliminary Analysis

The preliminary analysis provides the basis for: (1) estimating if the building has adequate capacity to withstand the specified structural criteria which may include code compliance; (2) property damage assessment; and (3) identifying structural deficiencies in the building with special reference to applicable minimum life and public safety standards and codes. Engineering judgment and experience in conjunction with simplified analysis techniques are required to determine rational structural demands and capacities for the critical members of the building. Environmental effects on the structure should be considered as part of the assessment. The following subsections provide discussion of key aspects of the preliminary analysis.

2.3.3.1 Loading and Performance Criteria

The engineer should establish and review the loading and performance criteria to be used in the

GUIDELINE FOR STRUCTURAL CONDITION ASSESSMENT OF EXISTING BUILDINGS

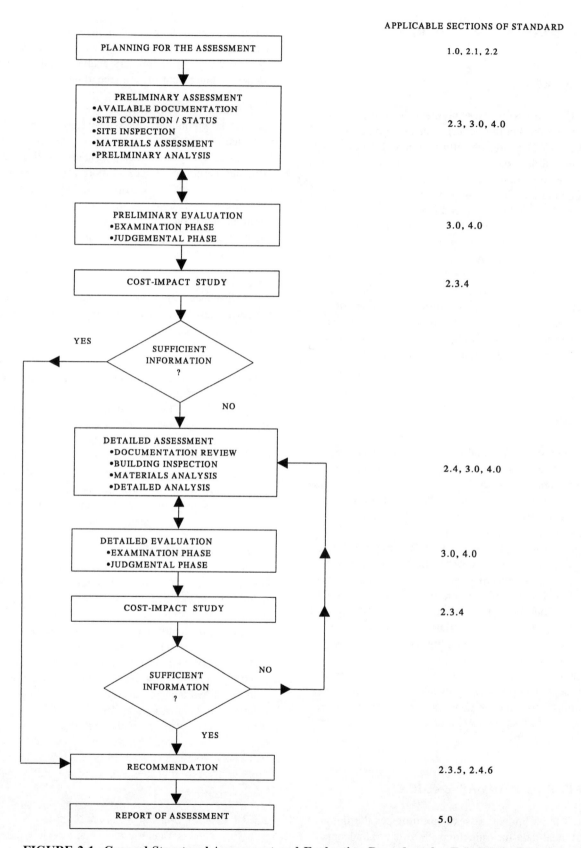

FIGURE 2-1. General Structural Assessment and Evaluation Procedure for Existing Buildings

preliminary analysis, such as current national, state, or local codes; federal regulations; or owner-prescribed requirements. Moreover, the engineer should attempt to become familiar with the design code and detailing practices in effect when the building was constructed. The performance criteria should establish the predominant performance requirement imposed on the building such as basic life safety of the occupants or essential operational capability.

2.3.3.2 Structural Components

Based upon the review of the available documentation, results of the on-site inspection, and the loading and performance criteria; the engineer should identify the primary vertical and lateral load resisting systems in the building and the primary vertical and lateral force paths that transfer the forces to the foundation. Structural members and connections in the structural system in the vertical, transverse, and longitudinal directions should be identified, and the physical properties and details for these members and connections determined. Critical members and connections are those whose failure would seriously reduce the capacity of the structure to resist the applied forces.

2.3.3.3 Material Properties

The existing construction drawings may contain information regarding the material properties used in the design. If not available, the strength of the materials can be estimated based upon the design criteria and the types of structural materials commonly used at the time of design and construction. An estimate of the material properties is generally adequate for the preliminary assessment. If the preliminary assessment reveals that certain material properties have a significant impact on the building's structural adequacy, in situ testing may be justified. Both nondestructive and destructive procedures may be necessary in order to establish more realistic properties for the detailed assessment. Section 3 provides a summary of the physical properties of various construction materials and tests that can be conducted to determine the physical properties.

2.3.3.4 Member Analysis

The critical members and connections in the structural system should be analyzed to determine their resistant capacities for moment, torsional, axial, and shear forces, and to compare their resistant capacities with demand capacities associated with the specified loading and performance criteria. Capacity reduction factors may be used to account for uncertainties such as detailing practices which are different from those currently recommended, the condition of the structure and material, and section properties of the member.

2.3.3.5 Nonstructural Components

If the existing building has nonstructural components which may significantly contribute to its structural resistance, the components should be included in the preliminary analysis. The interaction of the nonstructural components with the structural framing systems or the anchorage and attachment of these nonstructural components must be considered by the engineer. The integrity of nonstructural components may also be an important consideration.

2.3.3.6 Structural Evaluation

The engineer shall integrate the information and data regarding the existing building with the results from the analyses of critical components and connections to determine the overall structural condition of the building. Section 4 provides information on the evaluation of various construction materials. If the structural condition is marginal, detailed evaluation may be required. If the structural condition indicates significant deficiencies, a rehabilitation study may be warranted or a recommendation can be made to phase the building out of use.

2.3.4 Cost-Impact Study

A cost-impact study may be conducted for a building to be assessed. The cost of structural rehabilitation is subject to many factors; however, the cost for certain types of structural strengthening work can often be roughly estimated. Such an estimate can form the basis for an initial decision regarding the economic feasibility of the rehabilitation project. If results of this initial evaluation indicate that strengthening is economically feasible, a more detailed analysis may be required. A detailed analysis may provide a reasonably accurate estimate of the total costs and determine the economic feasibility of the overall project, including such factors as disruption costs of temporarily vacating the building. If the cost of rehabilitation is determined to be prohibitive, alternate uses of the building may be established or a plan made for phasing out its use.

2.3.5 Preliminary Condition Evaluation and Recommendations

The results of the preliminary assessment shall be summarized in a report (see Section 5). The report will include the following items: 1) description of loading and performance criteria considered; 2) description of building; 3) description of preliminary evaluation process; 4) discussion of preliminary findings; and 5) recommendations concerning the need for particular actions such as conducting a detailed assessment.

2.4 DETAILED ASSESSMENT

A detailed assessment shall be performed on the building being analyzed as a result of the findings of the preliminary assessment or as directed by mandatory actions or by the client. The purposes of the detailed assessment are: 1) to determine if the building satisfies the required performance criteria or if it requires rehabilitation; and 2) if the building requires rehabilitation, to identify its deficiencies and recommend alternatives for rehabilitation. The detailed assessment process is similar to that used in the preliminary assessment except that it is done in greater detail and with more accuracy in order to increase the reliability of the resulting recommendations.

2.4.1 Document Review

Documents governing or prepared for the original design and construction, rehabilitation, alteration, repair of, or addition to the building shall be reviewed, if available. This documentation may include drawings, specifications, mill test reports, calculations, geotechnical investigations and foundation reports, standards and codes in effect at the time of construction, construction records, operations and maintenance data, records of alterations made to the building, information on original materials and components in the building, and site records. Possible sources of building documentation include: 1) building owner and on-site staff interviews; 2) local building department records; 3) interviews with design professionals (architects and engineers) and contractors associated with the original construction or subsequent modification of the structure (or their successor firms); files of the original or successor design firms and contractors; 4) standards and local codes in effect at the time of construction; 5) files of insurance companies that may maintain drawings or records of their insured buildings; 6) professional publications dated within approximately one year of the building's construction or subsequent modification(s); 7) local library, historical society, and newspaper files; 8) manufacturer data on materials, components and systems; 9) weather data; and 10) topographical information.

2.4.1.1 Building Features

Available design drawings and calculations should be carefully reviewed to identify irregularities in the building's configuration or other features which will have a major influence on its behavior. Irregularities in the plan and vertical configuration may significantly affect the behavior of a building under certain loading conditions and the engineer shall carefully assess the impact of such irregularities. Common irregularities include: mass and resistance eccentricity; common geometric irregularities such as +, H, U, E, T, and L-shapes; unsymmetrical geometry with respect to the vertical axis of the building; setbacks at one or more levels; discontinuities in diaphragm stiffness; and discontinuous shear walls. The support of large equipment loads by the structure should also be considered.

2.4.1.2 Structural System

The type and continuity of the existing structural system has a major effect on the performance of the building. Flexible buildings can undergo large deformations and relative movements; therefore, potential damage to nonstructural elements and systems is greater. Stiff buildings will undergo smaller deformations, and the potential damage to nonstructural elements and systems is less. Rigid diaphragms tend to reduce the effects of unsymmetrical form. The most important factor, however, is the degree to which the building is tied together as a unit with all connections having adequate capacity for load transfer by shear, tension, compression, or moment as applicable. The structural engineer shall carefully evaluate the integrity of the structural system. This evaluation should reflect the engineer's assessment of the redundancies and discontinuities in the building structural system, the structural integrity of connections that tie the building together, the toughness and ductility of the construction material, the regularity of the building both in plan and vertical configurations, and the effects of unusual structural features. Most of the factors cited are difficult to quantify; therefore, it is essential that an experienced structural engineer perform the structural assessment. Common problems include discontinuities with the vertical and lateral load resisting systems, lack of ties and continuity, lack of connector elements, and nonredundant systems.

2.4.2 Building Inspection

After the design and construction documents and the result of any previous inspections, evaluations, or testing reports have been reviewed, the engineer and any required consultants should inspect the building to verify that the existing structural system and components meet the intent of the design and determine if there are any significant variations. The on-site inspection also affords an opportunity to inspect the nonstructural components which may be hazardous or present specific problems affecting the structural system. Testing or special inspections should be made when there are insufficient data to make this determination or when as-built conditions are suspect.

2.4.3 Detailed Assessment

While the preliminary analysis focused on critical members and connections, the detailed assessment will, in general, encompass the structural systems and their interactions as a total structural system.

2.4.3.1 Loading and Performance Criteria

The specified loading and performance criteria applicable to the detailed assessment should be reviewed for completeness. Acceptable performance criteria should be established for each potential mode of behavior, such as elastic and post-yield levels, and should include capacity reduction factors, drift and deflection limitations, and damping values. The performance criteria should be used to establish the predominant performance requirement imposed on the building, such as the basic life safety of the occupants or essential operational capability.

2.4.3.2 Critical System/Members

Critical structural components including members and connections comprising the structural system should be identified based upon the document review, the building inspection, the results of previous inspections, and evaluation and tests, if applicable. Components may also be considered critical if their current condition is adversely affected by corrosion, settlement, deterioration, etc.

2.4.3.3 Material Properties and Detailing Practices

A testing program may be established to find those material properties and detailing practices that do not conform to currently accepted practices. Test results may make the difference between accepting a building in an as-built condition or requiring a costly modification. Structural capacities of existing materials will be determined in accordance with the criteria and testing requirements in Section 3.

2.4.3.4 Capacities of Existing Systems/Members

The bending moment, shear, torsional shear, and axial load capacities of the critical structural systems, including members and connections comprising such systems, should be determined, preferably by ultimate strength methods. The material and section properties data for an existing building should be interpreted carefully, because the detailing may not conform with new construction requirements. Interaction curves may be required to establish the capacity of columns and walls.

2.4.3.5 Required Capacities of Systems/Members

The specified vertical and lateral loading criteria should be used to determine the required capacities of the critical structural systems, including members and connections, that must resist the specified loads and satisfy the drift and deflection limitations.

2.4.3.6 Nonstructural Components

If appropriate, the nonstructural components should be evaluated to ensure that they resist the prescribed forces and deformations. The effect of nonstructural components on the building's performance should also be considered. All deficiencies should be noted.

2.4.3.7 Structural Evaluation

The building shall be evaluated by means of an actual/required capacity comparison. If the building meets the performance criteria and does not have any deficiencies, the structural condition is adequate. If it does not conform to the performance criteria, a more sophisticated analysis procedure to evaluate the building may be used. If the building still does not conform with the acceptable performance criteria a realistic recommendation regarding its disposition should be formulated.

2.4.4 Detailed Findings

The detailed findings should indicate whether the existing building conforms with the performance criteria and should state the basis for this finding. If the building does not conform with the performance criteria, the deficiencies should be clearly identified and explained. Alternate concepts for correcting the deficiency or upgrading the building to conform with the acceptable performance criteria should be identified and described.

2.4.5 Cost–Impact Study

A cost study as described in Section 2.3.4 may be required for the detailed assessment.

2.4.6 Recommendations

The results of the detailed assessment shall be summarized in a report (see Section 5) with a recommendation as to the appropriate course of action. This could include acceptance of the building as is, rehabilitating the building to correct the deficiencies identified, changing the use of the building, or phasing the building out of service.

3.0 STRUCTURAL MATERIALS ASSESSMENT

3.1 PURPOSE AND SCOPE

State-of-the-art information is presented on practices and procedures for assessing the condition of selected materials in buildings. These practices and methods include visual inspection, nondestructive evaluation (NDE) tests, and destructive tests including both field and laboratory procedures. The practices and methods included in this Section and Section 4 are described in general terms with reference to documents which provide additional information. The information obtained from using the described practices and methods may assist in selection of additional tests to be used in the evaluation process. Other appropriate practices and tests should not be excluded from consideration.

3.2 CONDITION ASSESSMENT OF STRUCTURAL CONCRETE

3.2.1 Introduction

Structural concrete includes unreinforced and reinforced concrete (both cast-in-place and precast), prestressed concrete (pre-tensioned or post-tensioned), or combinations thereof. Condition assessment of concrete may require determinations of strength and quality. Proper assessment and subsequent evaluation (see Section 4.2) may reveal the concrete's ability to sustain loads and environmental conditions to which it is being or will be subjected.

The condition of a concrete structure may be determined by visual examination in conjunction with computational methods and testing. Major properties and physical conditions are discussed in Section 3.2.2. In addition, a guide to selecting test methods which can be used to evaluate these properties and physical conditions is provided in Section 3.7; and Tables 3.7.1, 3.7.2, 3.7.3, 3.7.4 and 3.7.5. Basic information for conducting an engineering analysis of structural concrete structures is found in ACI 318-95, Building Code Requirements for Reinforced Concrete and Commentary.

3.2.2 Properties and Physical Conditions of Concrete

The determination of the properties and physical condition of structural concrete may be required to provide an indication of its condition. Observed problems in the structure may indicate areas requiring investigation. Commonly measured properties and physical conditions are defined and discussed in this Section.

3.2.2.1 Properties of Concrete

Absorption: The increase in mass of concrete resulting from the penetration of water into the pores. Usually measured by submerging a concrete specimen in water. Considered to be a predictor of the durability of concrete.

Acidity: Concrete is chemically basic, having a pH of about 13, and therefore is attacked by acids (pH values less than 7).

- The American Concrete Institute (ACI 5.15.1R-93, 1993) has prepared a list of over 250 chemicals documenting their effect on concrete.

Air Content: The volume of air voids in concrete, usually expressed as a percentage of total volume of the paste, mortar, or concrete.

- Concrete that exhibits damage from freezing and thawing will invariably have an air content below the percentage recommended (ACI 201.1R-92, 1992).

Carbonation: Reaction between calcium hydroxide or oxide in cement, mortar or concrete with carbon dioxide to form calcium carbonate which can reduce the pH of the concrete.

Cement Content: Mass of cement in concrete, an important factor in determination of the heat generated during hydration and hence the susceptibility of the concrete to shrinkage.

- The durability of concrete depends on the impermeability of the cement paste. One way to insure the necessary degree of impermeability of the paste is to specify a minimum cement content for a mixture (Mather, 1989).

Chemical Content: Presence or amount of foreign compounds in the concrete.
- Most foreign compounds in concrete result in a reduced compressive strength. An example is calcium chloride which has often been used beneficially as an accelerator but excess amounts can be detrimental to reinforced concrete by inducing corrosion of reinforcing steel (ACI 318-95, 1995).

Chloride Content: Mass of free chloride in the concrete, a significant measure of active corrosion of reinforcing steel.
- Reinforced concrete exposed to chloride ions rapidly deteriorates once the chloride content exceeds the corrosion threshold limit. The diffusion of the chloride ion through concrete follows Fick's law, which is a function of time, the chloride concentration at the surface, and a diffusion constant (Cady and Weyers, 1983; Weyers and Smith, 1989).

Compressive Strength: Measured maximum resistance to axial compressive loading expressed as force per unit cross-sectional area.
- The compressive strength of concrete is one of its most important properties, not only because strength is how concrete is designated, but also because this property is directly and indirectly related to many other concrete properties. These include density, durability, modulus of elasticity, modulus of rupture, permeability and soundness. Concrete strength is determined by a standardized testing procedure. The resulting strength is an arbitrary value that may not have much relationship to the concrete in the structure. However, it is by this value that concrete is designated, designed and purchased so that any investigation of concrete strength to evaluate its design strength should attempt to correlate investigative data of tests on specimens taken from the hardened concrete to the standardized test. This means that results must be corrected for sample size differences (Section 3.7, Table 3.7.1, ASTM C42-90), for maturity (Section 3.7, Table 3.7.1, ASTM C1074-93; Plowman, 1956), and statistically (Section 3.7, Table 3.7.1, ACI 214.3R-88). The statistical correction is difficult because a minimum of 15 tests are required to determine a standard deviation.

Creep: Increase of strain under sustained loading.
- 90% of creep occurs in the first 2 years of life (ACI 318-95, 1995). It can generally be assumed all creep deflection has occurred in older concrete structures. However, additional creep can occur if new loads are applied to the structure.

Density: Mass per unit of volume.
- Concrete with an air dry mass not exceeding 115 lb/ft^3 (1,850 kg/m^3) is defined as a structural lightweight concrete (ASTM C567-91, 1991).

Elongation: Increase in length of a specimen as a percentage of the original length.

Modulus of Elasticity: Ratio of normal stress to corresponding strain for tensile or compressive stresses below the proportional limit of the material.
- Because of the relationship between concrete density and concrete compressive strength with the modulus of elasticity, the determination of density and compressive strength is usually sufficient to approximate the modulus of elasticity.

Modulus of Rupture: Determination of flexural strength of plain concrete; used as a measure of development of tension cracks under flexure.
- See comment under Tensile Strength

Moisture Content: Percent by mass of free water; an important factor affecting concrete volume changes.
- Concrete volume changes in long span structures may be significant and might have to be considered in a structural investigation.

Permeability: Measurement of the capacity for movement of water through concrete.

Proportions of Aggregate: Respective proportions of the fine and coarse aggregate in concrete by mass.

Pullout Strength: Strength determined by measuring the maximum force required to pull an insert from the hardened concrete.

Resistance to Freezing and Thawing: Used to determine the effect of properties of concrete on the resistance of concrete to freezing-and-thawing cycles; may be used as a measure of concrete durability.

Soundness: Freedom of a solid from cracks, flaws, fissures, or variations from an accepted standard.

Splitting Tensile Strength: Indirect method of applying tension to a specimen to cause failure by splitting; used both as a measure of diagonal tension capacity and of tensile strength.

Tensile Strength: Maximum unit stress which a material is capable of resisting under axial tensile loading, based on the cross-sectional area of the specimen before loading.
- Because of the difficulty of testing concrete in tension, the splitting tensile strength and the

modulus of rupture are both used as indirect measures of tensile strength. However, direct procedures are available (Hughes and Chapman, 1966; Evans and Marathe, 1968; Gopalaratnam and Shah, 1985).

Uniformity of Mix: Homogeneity of mix; cement, aggregate type, and dispersal of ingredients throughout the concrete.

Water-Cement Ratio: Ratio by mass of mixing water to cement; a significant indicator of concrete strength and durability.

3.2.2.2 Properties of Reinforcing Steel, Pre-Tensioning Steel, and Post-Tensioning Steel

Bend Test: Requirement for steel wires and bars used for concrete reinforcement to establish their ability to be bent around a pin without cracking.

Breaking Strength: Measure of the strength of steel strand fabricated from individual wires of varying tensile strengths. It is the equivalent tensile strength of the steel strand.

Carbon Content: Indicator of the property of steel bars used for concrete reinforcement to establish the weldability of the steel.

Chemical Composition: Indicators of properties of metals which can be used to establish corrosion resistance and mechanical properties.

Coating Properties: The requirements for adhesion, continuity, and thickness of epoxy on epoxy-coated wires and bars used for concrete reinforcement.

Deformation Requirements: Required spacing, height, and other physical features of the deformations on steel bars used for concrete reinforcement to meet bond strength requirement.

Elongation: Increase in gauge length of the tension test specimen measured after fracture as a percentage of the original length; a measure of ductility.

Reduction of Area: Decrease in the area of the fractured section of a tensile test specimen as a percentage of the original area.

Strength of Connections: A strength requirement of connections in welded or clipped deformed bar mats used for concrete reinforcement.

Tensile Strength: Stress calculated from the maximum load sustained by a metal specimen during a tension test divided by the original cross-sectional area of the specimen.

Weld Shear Strength: Requirement for steel welded wire fabric used for concrete reinforcement to substantiate the bonding and anchorage value of the wires.

Yield Strength: Stress at which significant increase in strain occurs without such increase in stress.

3.2.2.3 Properties of Connections

Connections in cast-in-place structural concrete (unreinforced, reinforced, pre-tensioned, post-tensioned, or combinations thereof) are integral with members; consequently there are no separate connections to be assessed. Connections in precast structural concrete construction (unreinforced, reinforced, pre-tensioned, post-tensioned, or combinations thereof) are always of different materials (e.g., different types of bearings or embedded or attached steel shapes which are attached by connectors) and reference is made to the appropriate material section of this Standard for guidelines on assessment of such connections.

3.2.2.4 Properties of Connectors

Connectors are used to make the connections in precast structural concrete. These connectors are always the same as used for connectors of metals and reference is made to Section 3.3.2.3 for the appropriate properties.

3.2.2.5 Physical Conditions of Concrete

Alkali-Carbonate Reaction: The reaction between alkalies (sodium and potassium) in portland cement and certain carboniferous rocks or minerals, such as calcitic dolomites and dolomitic limestones; products of the reaction may cause abnormal expansion and cracking of concrete in service. The reaction may occur internally between the cement and the aggregates in the concrete, or externally between the cement and carboniferous rocks or minerals in the surrounding environment.

- Problems associated with dolomite limestones are uncommon and most limestone aggregates used in concrete have good performance records and are innocuous. Reactive dolomite limestones are very fine grained, dense and frequently contain illite and chlorite clay minerals (Section 4.9.1, Hobbs, 1988).

Alkali-Silica Reaction: Reaction between the alkalies (sodium and potassium) in portland cement and certain siliceous rocks or minerals, such as opaline chert and acidic volcanic glass present in some aggregates; products of the reaction may cause abnormal expansion and cracking of concrete in service. The reaction usually occurs internally between the cement and the aggregate in the concrete; however, it could occur externally between the cement and sili-

ceous rocks and minerals in the surrounding environment (such an occurrence is uncommon).
- Alkali-silica reaction in the short term is usually controlled by use of nonreactive aggregates and use of low-alkali portland cement. However, even with these precautions, concrete exposed to salt water or a salt-water atmosphere over a long period of time can absorb sodium ions leading to eventual alkali-silica reaction. This problem may be controlled by use of appropriate amounts of pozzolans in the concrete (Gerwick, 1991).
- Alkali-silica reaction can result in cracking of the concrete surface in a mapping pattern [see Fig. 3.2.2.5(1)].

Bleeding Channels: Passageways left by escape of excess water during drying and settlement; bleeding channels weaken the concrete and allow water, salts, and other chemicals to attack the concrete interior.
- It has been suggested that a greater bleeding capacity of the concrete decreases the plastic shrinkage and that shrinkage cracking takes place when the rate of evaporation exceeds the rate at which bleeding water rises to the surface (ACI Monograph No. 6, 1971).

Carbonation: Results between calcium hydroxide in hardened concrete and atmospheric carbon dioxide which can induce the corrosion of reinforcing steel.

Cement-Aggregate Reaction: Reactions between cement and constituents of aggregate, or other than alkali-silica and alkali-carbonate, which include hydration of anhydrous sulfates, rehydration of zeolites, wetting of clays and reactions involving solubility, oxidation, sulfates and sulfites. The products of these internal reactions may cause abnormal expansion and cracking of concrete in service.
- Recent research indicates that the cement-aggregate reaction is mainly a reaction between the alkalies in the cement that produce high pH and abundant hydroxyl and siliceous constituents of the aggregates. However, the field performance of concretes made with reactive sand-gravels does not correlate well with cement alkali content.

Chemical Degradation: Degradation of concrete due to chemical attack, such as sulfate attack, acid attack, and alkali-aggregate reactions.
- See Section 3.2.2.1, Acidity

Chloride Attack: Chlorides and nitrates of ammonium, magnesium, aluminum, and iron attack concrete, with those of ammonium being the most harmful. Sodium chloride is chemically harmless to concrete, but is a major contributor to corrosion of reinforcing steel.
- The primary modes of penetration of chloride through concrete is through cracks and by diffusion (Cady and Weyers, 1983).

FIGURE 3.2.2.5(1). Mapping Cracking

Contaminated Aggregate: Presence of impurities which either may interfere with the chemical reactions of hydration, or induce chemical degradation of concrete.
- In regions where water is scarce, inadequate washing of the aggregate to remove organic materials and environmental deposits commonly results in contaminated aggregates.

Contaminated Mixing Water: Impurities in the mixing water may interfere with chemical reactions of hydration, may adversely affect the strength of concrete, cause staining of the surface, and may also lead to corrosion of the reinforcement.

Cracking: Separation of concrete into parts characterized by length, direction, width, and depth, and whether the crack is active or passive. Passive cracks may be caused by construction errors, shrinkage, variations in internal temperature, or shock waves. Active cracks may be caused by variations in atmospheric or internal temperature, absorption of moisture, reinforcement corrosion, chemical reactions, settlement, or various loading conditions.
- Research evidence strongly suggests that the bond between mortar and aggregate is the weak link in the heterogeneous concrete system. This relative weakness of the interface between cement paste or mortar and aggregates is the source of microcracking. The onset of cracking arises mainly from spread of microcracks along the boundaries of the coarse aggregate (Shah and Winter, 1968; ACI Monograph No. 6, 1971).

Cross-Sectional Properties: Dimensions and other geometric properties of the structural components.

Delamination: A horizontal splitting, cracking, or separation of a concrete member in a plane roughly parallel to, and generally near the surface; found most frequently in bridge decks and parking deck slabs and caused by the corrosion of reinforcing steel. It is not readily apparent without testing such as soniscoping, chain-dragging, or tapping.
- Delamination usually occurs in the plane of the reinforcing steel as shown in Fig. 3.2.2.5(2).

Deterioration: Impairment of usefulness of concrete.

Discoloration: Departure of color from that which is normal or desired; usually caused by stains or chemical changes.

Disintegration: Deterioration into small fragments or particles due to factors such as chemical attack, weathering, and erosion.

Distortion: Warping or deforming of concrete due to factors such as overloading, poor design, ground movement, and expansion.

Efflorescence: Deposit of white salts, usually calcium carbonate, on the concrete surface. Fig. 3.2.2.5(3) shows cracking due to alkali-silica reaction and efflorescence on the face of the wall.

Erosion: Progressive surface disintegration of concrete by the abrasive or cavitation action of gases, fluids, or solids in motion.

FIGURE 3.2.2.5(2). Delamination of Concrete Caused by Corrosion of Reinforcing Steel

FIGURE 3.2.2.5(3). Cracking of Concrete Due to Alkali-Silica Reaction and Efforescence on the Face of the Wall

Freeze Thaw Damage: Surface damage to concrete resulting from the cycle actions of freeze thaw.
- A close-up view of the results of freeze thaw damage is shown in Fig. 3.2.2.5(4).
- Concrete may crack due to internal expansion caused by freezing of trapped water in the capillary cavities in hardened cement paste. This will not happen if the cement paste is provided with a system of entrained air voids so that the bubble spacing factor of paste is 0.008 in. (0.2 mm) or less (Mather, 1968).
- Also, certain aggregates are susceptible to frost attack (ACI 201-92, 1992).

Honeycomb: Voids left in concrete due to lack of adequate vibration, inappropriately low slump, or congestion of the reinforcing steel.

Leaching: Dissolution and cement constituents, often alkalies, from the interior of the concrete to the surface.

Popouts: The breaking away of small portions of a concrete surface due to internal pressure which leaves a shallow, typically conical depression. Popouts may be caused by reinforcing steel corrosion, cement-aggregate reactions, or internal ice crystal formation.
- Popouts range in diameter from 1 in. (25 mm) to 4 in. (100 mm) and depth from 0.5 in. (13 mm) to 2 in. (50 mm). Popouts are measured by the number per ft^2 (m^2) (SHRP-P-338, 1993).
- A typical popout is shown in Fig. 3.2.2.5(5).

Scaling: Local flaking or peeling away of the cement matrix near-surface portion of hardened concrete or mortar. Scaling is a disintegration process.

FIGURE 3.2.2.5(4). Freeze Thaw Damage to Concrete Wall

- Scaling is shown in Fig. 3.2.2.5(6).

Spalling: Detachment of fragments (spalls), usually in the shape of flakes, from a concrete mass. Variations in internal temperature, corrosion of reinforcement bars, chemical reactions, weathering, and poor design can cause active spalling. Shock waves or a single incident of varying internal temperature can cause passive spalling. Passive spalls may simply be repaired, but active spalls are warnings of a greater, possibly dangerous problem.

Stratification: The segregation of over-wet or over-vibrated concrete into horizontal layers with increasingly lighter material toward the top.

Structural Performance: Excessive cracking or other indications of structural distress due to excessive applied loads, or deformations due to damage from fire or other external means.

Sulfate Attack: Reaction between sulfate solutions (resulting from, for instance, alkali magnesium and sodium sulfates in groundwater and seawater) and hardened cement paste (the sulfate reacting with the calcium aluminate in the cement paste). The products of the reaction lead to expansion and cracking of the concrete.

- Concrete may crack due to internal expansion resulting from reaction of sulfates in solution with aluminate hydrates of the cement. If the amount of aluminate hydrate is sufficiently large, such cracking may take place due only to interaction with moisture from the environment if the source of the sulfates were in the cement or the aggregate. This, however, is not the normal condition; usually the source of the sulfates comes from the environment (Mather, 1968).

Uniformity of Concrete: Degree or consistency of the properties of the concrete from one part of the structure to another.

Unsound Cement: Presence of compounds in the cement that results in appreciable expansion under conditions of restraint causing disruption of the hardened cement paste. Such expansion may take place due to delayed or slow hydration of free lime, magnesia, and calcium sulphate.

- Concrete may crack due to internal expansion caused by reaction of moisture with unhydrated calcium oxide or magnesium oxide that was introduced into the concrete as part of the cement (Mather, 1968).

Unsound Concrete: Concrete which has undergone deterioration or disintegration during service exposure.

3.2.2.6 Physical Conditions of Reinforcing Steel, Pre-Tensioning Steel, and Post-Tensioning Steel

Anchorage: In reinforced concrete, an extension of the reinforcement, either straight or in a bent

FIGURE 3.2.2.5(5). Popout

FIGURE 3.2.2.5(6). Scaling of Concrete

shape, used to transfer the force in the reinforcement to a support. In pre-tensioned and post-tensioned concrete, a separate device used for the same purpose.

Corrosion: Disintegrating or deterioration of reinforcement by electrolysis or chemical attack.

Cover: The least distance between the surface of the reinforcement and the outer surface of the concrete.

Cross-Sectional Properties: Actual dimensions and other geometric properties of reinforcement.

Dimensional Location: Dimensional location (spacing, position, length) of reinforcement.

Exposure: Condition where reinforcement has been visibly exposed by loss of its protective cover.

Shape: Physical configuration of reinforcement; its bent shape which may be either two or three dimensional.

3.2.2.7 Physical Conditions of Connections

For cast-in-place structural concrete (unreinforced, reinforced, pre-tensioned, or combinations thereof) the connections are integral with the members and there are no separate connections to be assessed. For precast concrete (unreinforced, reinforced, pre-tensioned, post-tensioned or combinations thereof) the connections are always of different materials (e.g., different types of bearings or embedded or attached steel shapes which are attached by connectors) and reference is made to the appropriate material section in this Standard for assessment of the appropriate physical conditions.

3.2.2.8 Physical Conditions of Connectors

Connectors are used to make the connections in precast structural concrete. These connectors are always the same as used for connectors of metals and reference is made to Section 3.3.2.6 for the appropriate physical conditions.

3.2.3 Test Methods

Visual, nondestructive, and destructive tests are described in Section 3.7. In Tables 3.7.1, 3.7.2, 3.7.3, 3.7.4 and 3.7.5, each test is briefly explained and its requirements, advantages, and limitations discussed.

3.2.4 Combinations of Test Methods

It may be necessary to use a combination of methods for satisfactorily predicting the strength or quality of concrete. For example, in predicting the compressive strength of in situ concrete more accurately, two or even three different tests may be performed, and their results combined. A typical combination is the ultrasonic pulse velocity method in conjunction with the rebound hammer.

3.2.5 References

References cited in Section 3.2, unless otherwise indicated, are included in Section 3.6. Additional References are to be found at the end of Section 4 and after each of the tables in Section 3.7.

3.3 CONDITION ASSESSMENT OF METALS

3.3.1 Introduction

Metals commonly found in existing buildings for structural purposes are cast iron, wrought iron, steel, and aluminum alloy. Cast iron and wrought iron are very rarely used today as construction metals. However, cast iron and wrought iron members can be found in many buildings built in the 19th and early 20th centuries. Metallic components and systems carry the applied loads in bending, shear, torsion, axial tension or compression, or in combinations.

Condition of the structural metals in the buildings should be assessed for strength and stiffness by examining the material properties and physical conditions. Metals are vulnerable to fire if not properly protected. Cast iron, wrought iron, aluminum, and most steels are also vulnerable to corrosion. Due to the light mass of metals relative to their strength, large deflections and instability of metal elements can result.

Connections play a major role. Special attention should be paid to condition assessment of connections and connectors in metal structures.

Major properties and physical conditions of metals are discussed in Section 3.3.2. Section 3.7 and Tables 3.7.6, 3.7.7, and 3.7.8 provide guidance for selecting test methods which can be used to evaluate properties and the extent of defects. Basic information for conducting an engineering analysis of metal (steel) structures is found in AISC Manuals of Steel Construction (AISC-LRFD, 1994; AISC-ASD, 1989).

3.3.2 Properties and Physical Conditions of Metals

This Section discusses frequently measured chemical and mechanical properties and physical conditions of existing metal structures. Properties are measured to evaluate the strength and stiffness. Examination of the physical conditions of metals can be useful in evaluating strength, stiffness, degree of corrosion, and fire protection.

3.3.2.1 Properties of Metals

Carbon Content and Chemical Composition: Indicators of properties of metals which can be used to establish corrosion resistance and mechanical properties. Both carbon content and carbon equivalent content are needed to establish weldability of the metal.

Ductility: Mechanical property that is a measure of the ability to deform before cracking.

Elongation: Increase in gauge length of the tension test specimen measured after fracture as a percentage of the original length; a measure of ductility.

Fatigue Properties: Mechanical properties such as cyclic counting, endurance limit, *e-N* curve, hardness, and *S-N* curve that characterize resistance of the metals to fatigue crack initiation and propagation under low amplitude high frequency cyclic stresses or strains.

Fracture Properties: Mechanical properties such as Charpy number, fracture toughness, and nil ductility transition temperature that characterize resistance of the metals to brittle fracture, especially at low temperature.

Modulus of Elasticity: Ratio of normal stress to the corresponding strain in the linear elastic segment of stress-strain curve.

Reduction of Area: Decrease in the area of the fractured section of a tension test specimen as a percentage of the original area.

Tensile Strength: Stress calculated from the maximum load sustained by a metal specimen during a tension test divided by the original cross-sectional area of the specimen.

Yield Strength: Stress at which significant increase in strain occurs without such increase in stress or as defined by the appropriate standard specification for the metal.

- Relatively high strains (rates of loading) are used in mill tests relative to the slower rates of the essentially static loading which takes place in most structures. Since the yield point of steel increases with increasing strain rate, the values measured in mill tests are higher than those which are effective in the structure (Winter, 1961).

3.3.2.2 Properties of Connections

Connections are always made of metal so that the properties listed in Section 3.3.2.1 also apply for connections.

3.3.2.3 Properties of Connectors of Metals

Breaking Strength: A measure of the strength of wire strand and wire rope fabricated from individual wires of varying tensile strengths. It is the equivalent tensile strength of the wire strand or rope.

Chemical Composition: Indicators of the properties of bolts (including washers and nuts), rivets and studs which can be used to establish corrosion resistance and mechanical properties.

Coating Mass: A specified requirement for wire strand and rope, bolts and other metal connectors to insure an appropriate level of protection against corrosion.

Elongation: A requirement for studs, wire strands, and wire ropes measuring the increase in gauge length in a tension test specimen measured after fracture as a percentage of the original length; a measure of ductility.

Hardness: For rivets and threaded fasteners, sufficient hardness ensures that the surfaces of the fasteners have the capacity to transmit load by bearing; a specified requirement for rivets, nuts, and washers and sometimes for bolts and rods.

- Hardness is measured by a standard indicator using standard procedures as shown in Fig. 3.3.2.3(1).

Proof Load: A specified load without measured permanent set; applicable to bolts and rods.

Reduction in Area: Decrease in the area of the fractured section of a tensile test specimen as a percentage of the original area; a specified minimum requirement for studs to insure ductility.

Stress at 0.7% Extension: A mechanical requirement for wire strand and wire rope to establish the limiting load without fracturing of individual wires.

Tensile Strength: Stress calculated from the maximum load sustained by a metal connector during a tension test divided by the original cross-sectional area of the specimen; a required physical property for bolts, rods, studs, wire strand, and wire rope.

Tensile Strength of Filler Material: The same as Tensile Strength for Properties of Metals (Section 3.3.2.1) and a requirement of the physical property of electrodes used in welding.

Yield Strength of Filler Material: The same as Yield Strength for Properties of Metals (Section 3.3.2.1) and a required physical property of electrodes used in welding.

3.3.2.4 Physical Conditions of Metals

Bracing of Compression Elements and Members: Lack of necessary bracing can result in loss of stability, strength, and stiffness of main load carrying elements.

GUIDELINE FOR STRUCTURAL CONDITION ASSESSMENT OF EXISTING BUILDINGS

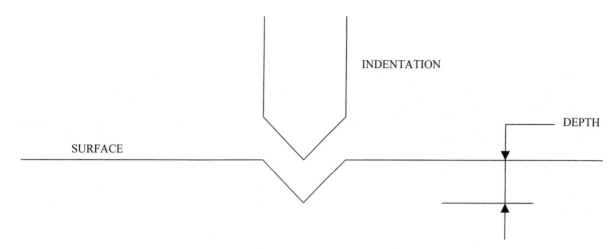

FIGURE 3.3.2.3(1). Hardness Test

Cross-Sectional Properties: Actual or nominal dimensions and other geometric properties of structural members.
- See Section 3.7, Table 3.7.8 (Ferris, 1953).

Deformations: Deformations of members in a metal structure which may be caused by fabrication or erection out-of-plumbness, lack of fit or slip at connections, settlement or failure of supports, overstressing, inadequate mechanical properties of materials used, inadequate bracing, change in temperatures, removed members or missing connectors, and torsional effects unaccounted for in the design.

Direct Chemical Attack: Deterioration of metals by attack of chemical solutions with either low or high acidity (pH).

Electrolytic or Electrochemical Corrosion: Oxidation of metal due to chemical reaction between the metal and oxygen in the environment; most common cause of deterioration of unprotected iron and steel products (rust).

Fatigue Cracking: Cracks that develop in ductile metals when the material is subjected to cyclic stresses. These cracks initially propagate slowly and, if detected in time, can be treated by taking remedial action. If the cracks are allowed to propagate unrestricted they frequently initiate brittle fracture.
- A procedure for identifying fatigue cracking by liquid penetrant is illustrated in Fig. 3.3.2.4(1).

Fire Protection: Strength and stiffness of metals may deteriorate significantly while subjected to elevated temperatures.

Fracture Cracking: Brittle cracks that take place with little or no preceding plastic deformation. Low temperature (possibly as high as room temperature), stress or strain concentrations (including microcracks and fatigue cracks), and metallurgical composition are important factors influencing fracture cracking. These types of cracks are often triggered by impact or sudden increase in load.
- Liquid penetrant also may be used to identify fracture cracking [see Fig. 3.3.2.4(1)]. Internal defects and cracks can be identified by use of either normal probe or angle probe ultrasonics as illustrated in Fig. 3.3.2.4(2).

Geometry of Structural Components: Actual or nominal dimensions of structural components.

Geometry of Structure: Actual or nominal dimensions defining geometry of structure.

FIGURE 3.3.2.4(1). Use of Liquid Penetrant to Identify Fatigue Cracks

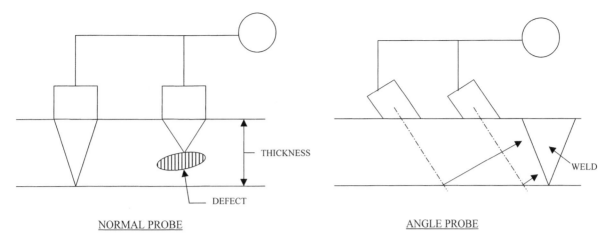

FIGURE 3.3.2.4(2). Use of Ultrasonic Probes

In Situ Stresses: The actual stresses that exist in a member representing the sum of manufacturing, fabrication and erection residual stresses and the stresses resulting from in situ loads and deformations.

Laminar Tearing: A planar separation that develops in flanges of rolled shapes and within thick plates near certain large welds as high weld shrinkage develops stresses across the plate thickness.

- Laminar tearing can be detected within a plate by using acoustic emission techniques as illustrated in Fig. 3.3.2.4(3).

Overall or Local Buckling: Metal members are relatively slender and usually have thin elements making them susceptible to overall or local buckling.

Overstressing: Subjecting metals to stresses in excess of their allowable stresses.

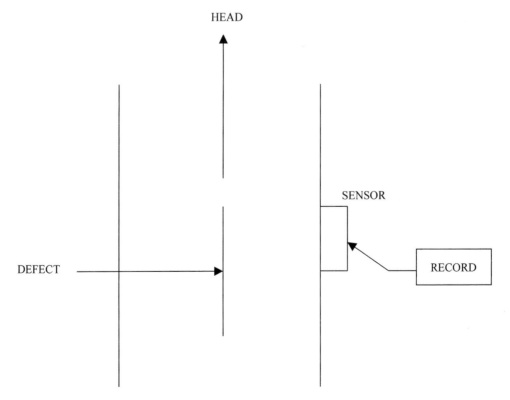

FIGURE 3.3.2.4(3). Use of Acoustic Emission

3.3.2.5 Physical Conditions of Connections

Connections are always made of metal so that the properties listed in Section 3.3.2.4 apply for connections.

3.3.2.6 Physical Conditions of Connectors

Condition: The physical state as determined by observation and with use of simple physical assistance such as cleaning, scraping, and sounding; can be used to establish existence of many of the physical conditions included herein.

Corrosion: Same as Electrolytic or Electrochemical Corrosion in Physical Condition of Metals (Section 3.3.2.4).

Cracks: Discontinuities in the weld filler material or in adjacent base metal in welded joints. The extent of cracks in welded joints may be ascertained by acid etching, magnetic particle inspection, or other means.

Cross-Sectional Properties: Actual dimensions of connectors.

Deformation: Elongation, bending, or twisting of bolts, rods, rivets, and studs over and above the elastic limit.

Dimensions: Actual diameter and length of bolts, rods, rivets, and studs.

Discontinuities: Either internal or external localized areas of interrupted filler material in welded joints that result from cracking, entrapped slag, inclusions, deoxidation products, gas pockets, or blow holes.

Length: The actual physical dimension of a length of weld or a length of wire strand or wire rope.

Location: The physical position of a weld relative to parts it connects.

Nonbinding Condition: Bolts or rods that are not fully restrained from movement by tight fitting nuts.

Porosity: The existence of gas pockets or blow holes in weld filler material.

Profile: The plot of the actual cross section through a weld connector. The profile can be flat, convex, or concave.

Size: The actual dimensions of a weld used to determine its capacity; size includes angle of bevel, depth of penetration, and root opening for butt welds. Fillet welds are sized by the length of the leg of the weld.

Slag Deposit: A discontinuity in a welded joint resulting from concentrations of impurities in the base metal as it is fused and from deposit of electrode coating material in the weld.

Slotted Holes for Movement: Slotted or oversized holes that permit connector movement primarily for thermal changes in the geometry of the structure. They are also used under certain conditions to control buildup of stresses in a connection.

Smoothness: The lack of abrupt and singular discontinuities in the weld profile.

Soundness: A physical condition of rivet connectors that can be determined by striking the rivet with a hammer and measuring by pitch of the resulting impact.

Tightness: The physical condition of bolts, rods, rivets, wire strand, and wire rope that indicates the connector fits snugly and generally under some positive compressive force.

Uniformity: A measure of the consistency of weld profile, weld size, weld dimensions, weld cross-sectional properties and conditions.

3.3.3 Test Methods

Visual, nondestructive, and destructive tests are used to establish properties and physical conditions of metal structures. Most common test procedures including requirements, advantages, and limitations are listed in Section 3.7, Tables 3.7.6, 3.7.7, and 3.7.8.

3.3.4 Combination of Test Methods

It may be necessary to use a combination of methods for satisfactorily predicting properties of metals (see Section 3.2.4).

3.3.5 References—Metals

References cited in Section 3.3 are included in Section 3.6. Additional references are to be found at the end of Section 4 and after each of the Tables in Section 3.7.

3.4 CONDITION ASSESSMENT OF MASONRY

3.4.1 Introduction

Masonry includes nonbearing walls (walls carrying their own mass as well as wind and seismic forces in some cases), bearing walls (such walls carry additional gravity loads above their own mass as well as, in some cases, in-plane and out-of-plane forces), beams, columns, piers, panels, curtain walls, floor arches, arches, and shells. Masonry is also used compositely with metals, concrete and wood.

Unlike other materials, masonry is a compound material. In its simplest forms, it is an assemblage of dry masonry units, or masonry units and mortar. In complex assemblages, different types of masonry units are used as well as different types of mortar. Masonry assemblages also can include steel reinforcing bars, grout, tie rods, embedded metals and wood, metal ties, anchors, and joint reinforcement. In some cases the masonry assemblage includes an air space, as in a cavity wall. The masonry units may be manufactured products or they may be natural building stones. For the condition assessment of other materials, reference is made to Section 3.2 for concrete and steel reinforcing bars, to Section 3.3 for metals, and to Section 3.5 for wood. This Section addresses the masonry units, mortar and grout of the masonry assemblage.

Basic information necessary for conducting an engineering analysis of masonry structures is found in ACI 530/ASCE 5/TMS 402, 1995 and ACI 530.1/ASCE 6/TMS 602, 1995.

3.4.2 Components of Masonry Assemblages

3.4.2.1 Manufactured Masonry Units

Manufactured masonry units include the following (see Section 3.7, Table 3.7.9 for references):

1. Structural Clay Load-Bearing Tile (ASTM C34, 1996).
2. Concrete Building Brick (ASTM C55, 1997).
3. Structural Clay Non-Load-Bearing Tile (ASTM C56, 1996).
4. Building Brick (Solid Masonry Units Made from Clay or Shale) (ASTM C62, 1997).
5. Calcium Silicate Face Brick (Sand-Lime Brick) (ASTM C73, 1997).
6. Load-Bearing Concrete Masonry Units (ASTM C90, 1997).
7. Ceramic Glazed Structural Clay Facing Tile, Facing Brick, and Solid Masonry Units. (ASTM C126, 1996).
8. Non-Load Bearing Concrete Masonry Units (ASTM C129, 1997).
9. Structural Clay Facing Tile (ASTM C212, 1996).
10. Facing Brick (Solid Masonry Units Made from Clay or Shale) (ASTM C216, 1997).
11. Chemical-Resistant Masonry Units (ASTM C279, 1995).
12. Hollow Brick (Hollow Masonry Units Made from Clay or Shale) (ASTM C652, 1997).
13. Prefabricated Masonry Panels (ASTM C901, 1993a).

3.4.2.2 Natural Building Stones

Natural building stone masonry units include the following (see Section 3.7, Table 3.7.9 for references):

1. Marble Building Stone (Exterior) (ASTM C503, 1996).
2. Limestone Building Stone (ASTM C568, 1996).
3. Granite Building Stone (ASTM C615, 1996).
4. Quartz-Based Dimension Building Stone (ASTM C616, 1996).
5. Slate Building Stone (ASTM Designation C629, 1996).

3.4.2.3 Mortar and Grout

Mortar and grout used in masonry include the following:

1. Mortar for Unit Masonry (ASTM C270, 1997).
2. Grout for Masonry (ASTM C476, 1995).
3. Extended Life Mortar for Unit Masonry (ASTM C1142, 1995).

3.4.2.4 Reinforcing Steel Bars

For reinforcing steel bars used in masonry, see Section 3.2.

3.4.2.5 Masonry Ties, Anchors and Joint Reinforcement

Masonry ties, anchors, and joint reinforcement are fabricated from metals and reference is made to Section 3.3.

3.4.2.6 Connections

Typically, mortar is used to bed or bond masonry units on or to each other. When mechanical connections are used, these are typically made of metals and reference is made to Section 3.3 for their condition assessment.

3.4.2.7 Connectors

In masonry construction, connectors may be considered as masonry ties, anchors, and joint reinforcement.

3.4.2.8 Other Materials

Masonry can be used compositely with concrete, metals, and wood. Reference is made to Sections 3.2, 3.3, and 3.5, respectively, for these materials.

3.4.3 Properties and Physical Conditions of Masonry

Properties of masonry have been separated into Properties of Masonry Units and Properties of Masonry Assemblages. However, physical conditions have not been separated since the physical conditions apply to both. Section 4.4 follows these approaches. Tables 3.7.9 and 3.7.10 combine tests on Masonry Units and Masonry Assemblages. However, the application of each test, either to units or assemblages, is indicated in the application column where appropriate.

3.4.3.1 Properties of Masonry Units

Masonry units may be removed from a structure for testing to determine their properties. Some masonry tests can be conducted in situ. Properties of masonry units include the following:

Absorption: The amount of water that a masonry unit (solid or hollow clay or concrete; or natural building stone) absorbs when immersed in either cold or boiling water for a stated length of time. It may be expressed as percentage of mass (clay masonry) or mass per unit volume (concrete masonry).

Adhesion: A property of prefaced masonry units. When viewed without magnification, no visible failure of the adhesion of the facing material to the masonry unit shall occur after the unit has been subjected to a standard compressive strength test.

Autoclave Crazing: Autoclave crazing is failure of the ceramic glazing on masonry units (structural clay facing tile, facing brick, and solid masonry units) when subjected to a standard steam pressure in an autoclave for a specified period.

Bulk Specific Gravity: A property of natural building stone equal to the mass of an oven-dried specimen divided by the difference between the mass of the soaked and surface-dried specimen in air and the mass of the soaked specimen in water.

Chemical Resistance: A property of the ceramic glazing on masonry units (structural clay facing tile, facing brick, solid masonry unit, and chemical resistant units) indicating its resistance to changes in color and texture when subjected to immersion in a specified acid solution for a specified length of time and temperature.

Compressive Strength: For a masonry unit, the maximum load divided by the gross area for brick units and the gross or net area for concrete masonry units as defined by the referenced standards when subjected to a standard strength test.

Dimensions: A characterization of concrete masonry units that includes standard measurements of width, height, length, face shell thickness, and web thickness.

Distortion: A measurement of warpage. Specified tolerances are required for hollow brick, facing brick, and ceramic glazed masonry units (see Warpage).

Drying Shrinkage: The change in linear dimension of a test specimen due to drying from a saturated condition to an equilibrium mass and length under specified accelerated drying conditions. Applicable to concrete masonry units.

Efflorescence: A deposit or encrustation of soluble salts, generally white, and most commonly consisting of calcium sulfate, that may form on the surface of a masonry unit when moisture moves through the unit and evaporates on the surface. Masonry units are rated as "not effloresced" and "effloresced."

Freezing and Thawing: The loss in mass as a percentage of the original mass of a dried specimen when subjected to standard cyclic testing of alternations of freezing and thawing.

Imperviousness: The resistance to staining of ceramic glazed masonry units when subjected to a standard test.

Initial Rate of Absorption (IRA): The mass of water absorbed when a brick or structural clay unit is partially immersed in water for one minute, expressed in $kg/m^2/min$ (also sometimes called suction).

- Flashed brick may exhibit significantly different IRA values from one bed surface to the other if one surface is flashed and the other is not. Molded bricks definitely have different IRA characteristics between bed surfaces (Bailey, Matthys, and Edwards, 1990).

Mass Determination: The mass per unit area of a specimen calculated by dividing the total mass, determined after a specified procedure of oven-drying and cooling of a unit, by the average projected area of the two faces of the unit as normally laid in a wall.

Modulus of Elasticity: The ratio, within the elastic limit of a material, of stress to corresponding strain under given loading conditions.

Modulus of Rupture: The value of the maximum tensile or compressive stress (whichever causes failure) in extreme fiber of a unit loaded to failure in bending computed from the flexure formula.

Moisture Content: The amount of moisture in a material determined under prescribed conditions and expressed as a percentage of the mass of the moist specimen; that is, the original mass comprising the dry substance plus any moisture present. For concrete masonry units, it is the percent of total absorption.

Moisture Expansion: An increase in dimension or bulk volume of some masonry units caused by reaction with water or water vapor.

- This reaction may occur in time at atmospheric temperature and pressure, but is expedited by exposure to water or water vapor at elevated temperatures and pressures.

Net Area: The net area of concrete masonry units is defined as the ratio of the net volume to the height of the unit. Net area calculations are based on values obtained in absorption and unit mass tests.

Opacity: The quality or state of a material that makes it impervious to the rays of light. A requirement for certain types of ceramic glazed masonry units.

Sampling: A method of selecting masonry units that are representative of the whole for testing. In general, five units are required to obtain a statistical sample for testing.

Saturation Coefficient: For brick masonry and structural clay tile, the ratio of the saturated mass of a specimen (after 24 h submersion in cold water) minus the dry mass of the specimen to the saturated mass of a specimen (after 5 h submersion in boiling water) minus the dry mass of the specimen.

Secant Modulus of Elasticity: The slope of the line connecting the origin and a given point on the stress-strain curve.

Size: A test for brick and structural clay tile masonry units that includes standard measurements of width, depth and length.

Solubility: The loss in mass as a percent of original mass as determined by a test for solubility in sulfuric acid. A requirement for chemical-resistant masonry units.

Splitting Tensile Strength: The failure load of a test specimen of a masonry unit when subjected to a specified compressive load applied to the unit that results in a tensile stress distributed over the height of the unit. Used as a measure of diagonal tension capacity and of tensile strength.

- The tensile strength of masonry units is an important characteristic affecting the capacity of concrete block masonry under various loading conditions. The tensile strength of the block is highly sensitive to the test technique. The strain gradient has a significant effect on block tensile strength (Hamid and Drysdale, 1980).

Surface Texture: The geometric irregularities in the surface of a masonry unit. Measurement of surface texture shall not include inherent structural irregularities unless these are the characteristics being measured.

Thermal Expansion: The increase in the dimensions or volume of a body due to change in temperature. The coefficient of thermal expansion is the change in unit length (or volume) accompanying a unit change in temperature, at a specified temperature.

Void Area in Cored Units: A test for brick and structural clay tile masonry units expressed as the unit's percentage of void area determined by a standard test procedure.

Warpage: A test for brick and structural clay tile masonry units that measures the maximum out-of-plane convexity or concavity to the nearest 1/32 in. (1 mm).

Water Absorption: The increase in mass of a test specimen after immersion in water under specified conditions of time and temperature, expressed as a percentage of its dry mass.

Young's Modulus: The ratio of normal stress to corresponding strain for tensile or compressive stresses less than the proportional limit of the material.

3.4.3.2 Properties of Masonry Assemblages

Masonry assemblages may be cut from a structure and tested as prisms or panels. Properties of masonry assemblages include the following:

Compressive Load: The maximum load divided by the appropriate bearing area that a masonry prism can sustain when subjected to a standard strength test.

- Strength and stress-strain characteristics of masonry prisms made with similar units differ markedly depending on the direction of compression and whether the prisms are hollow, solid, or grouted (Wong and Drysdale, 1983).

Concentrated Load: The maximum load that a masonry prism can sustain when subjected to a standard strength test. The concentrated load is applied transversely to the panel by a steel bar with a diameter of 1 in. (25.4 mm).

Deformability: The measurement of masonry assemblage deformability properties by use of testing equipment such as thin, bladder-like flatjack devices installed in cut mortar joints in the masonry and measuring load-deformation (stress-strain) properties.

Flexural Tensile Strength: The determination of the flexural bond strength of unreinforced masonry assemblages based on tests of simply supported beams either with third point loading or with uniform loading.

Modulus of Rupture: The value of the maximum tensile or compressive stress (whichever causes failure) in extreme fiber of a beam loaded to failure in bending computed from the flexure formula.

Secant Modulus of Elasticity: The slope of the line connecting the origin and a given point on the stress-strain curve. A property obtained in a strength test on a masonry prism.

Shear Strength: The determination of the diagonal tensile or shear strength of masonry assemblages by loading them in compression along one diagonal, thus causing a diagonal tension failure with the specimen splitting apart parallel to the direction of load.

Tensile Load: The maximum load that a prefabricated masonry panel can sustain when subjected to a standard tensile test.

Transverse Load: The maximum load that a prefabricated masonry panel can sustain when subjected to a standard transverse load test. The load is applied as two line loads located at the quarter points of the span being loaded.

3.4.3.3 Properties of Mortar and Grout

Mortar is a mixture of cementitious materials, aggregates, and water which, in its plastic state, is suitable for laying and bonding masonry units. Grout is a generic term reflecting a family of mortar and concrete mixtures with the specific mission of filling voids or cavities.

Samples of mortar and grout taken from structures are insufficient in size for compressive strength and splitting tensile strength testing. All physical properties of mortar and grout must be obtained through petrographic analysis. Properties of mortar and grout include the following:

Aggregate Characteristics: A visual examination can determine the composition, fabric, and condition of the aggregate. A microscopic examination can determine the shape, grading, distribution, texture, composition, rock types, alteration (degree and products), coating, rims, internal cracking, and contamination.

Constituent Portions: Chemical and analysis techniques are available for determining cement/lime/aggregate proportions.

Mortar and Grout Characteristics: A visual examination can determine cracking, color change, embedded items, and condition. A microscopic examination can determine existence of air entrainment, air voids (size, shape, and distribution), bleeding, segregation, aggregate-paste bond, fractures, alteration (degree, type, and reaction products) and nature and condition of surface treatments.

Paste Characteristics: A microscopic examination can determine color, hardness, porosity, carbonation, residual cement (distribution, particle size, abundance, and composition), mineral admixtures (size, abundance, and identification), compounds in hydrated cement and contamination (size, abundance, and identification).

- The primary function of masonry mortar is to provide for even bearing, bond the masonry units together, and provide resistance to water penetration (Hatzinikolas, Longworth, and Warwaruk, 1980).
- Grouted masonry testing shows grout compressive strength, per se, is a poor indicator of the interaction of grout in masonry (Isberner, 1980).

3.4.3.4 Physical Conditions of Masonry Units and Masonry Assemblages (Environmental)

The physical conditions of masonry that are influenced by environmental conditions (deicing salts, freezing and thawing, moisture, thermal changes, and weathering) include the following:

Air Leakage: The volume of air per minute at a measured temperature and pressure that flows perpendicularly through a masonry assemblage. Air movement is a significant factor affecting condensation.

Blistering: Swelling accompanied by rupturing of a thin uniform skin both across and parallel to the bedding plane, usually a condition found on sandstone, but also on granite.

Calcite Streak: A former fracture or parting (in limestone) that has been re-cemented and annealed by the deposition of obscure white or light-colored calcite.

Corrosion: The gradual wearing away of the exposed surface of masonry assemblages by external actions of chemicals or other forms of physical attack.

Coving: The hollowing out of a wall just above grade level, usually caused by standing rainwater or rainwater splash off the ground.

Cracking: Narrow fissures in masonry 0.0004 in. (0.01 mm) or larger in width.

Crazing: The formation of a pattern of tiny cracks or crackles in the glaze of glazed masonry units (structural clay facing tile, facing brick, and solid masonry units).

Crumbling: A condition indicative of a certain brittleness or tendency of the masonry to break up or dissolve.

Cryptoflorescence: A potentially harmful accumulation of soluble salts deposited under or just beneath the masonry surface as moisture in the masonry evaporates.

Delamination: A condition of stone in which the outer surface of the stone splits apart into laminae or thin layers and peels off the face of the stone.

Deterioration: The gradual degeneration or degradation of masonry resulting in a permanent impairment of the physical properties.

Dew Point: The temperature at which air becomes saturated with water vapor and below which moisture is likely to condense; varies with amount of moisture in the air.

Driving Rain Index: A measure of the likelihood of rain penetration determined by combining the annual precipitation with the average annual wind velocity.

Efflorescence: A deposit or encrustation of soluble salts, generally white and most commonly consisting of sulfates and chloride that may form on the surface of masonry when moisture moves through and evaporates on the masonry.

- All masonry materials may contribute to efflorescence. The occurrence of efflorescence should be used as a signal that all is not right either with the normal chemical reactions of the masonry materials or with the performance of the structure (Isberner, 1983).

Erosion: Wearing away of the surface, edges, corners of carved details of masonry slowly and usually by the mutual action of wind or windblown particles and water.

Exfoliation: Description of natural stone deterioration such as peeling, scaling, or flaking of the surface of stone in thin layers.

Friability: An inherent characteristic of some types of stone, particularly sandstone or limestone, which have a tendency to break up, crumble or powder easily.

Interstitial Condensation: An accumulation of moisture or water droplets that occurs within a porous masonry assemblage when moist air reaches its dew point, that is, when moist air is cooled to a point at which the water vapor (a gas) changes state into water (a liquid).

- Air leakage outward at the top of a building, especially at windows, can cause excessive condensation.

Peeling: A slipping away of the glaze surface or coating of masonry due to lack of adhesion or improper application of a surface treatment.

Pitting: The development or existence of small cavities in a masonry surface.

Rising Damp: The suction of ground water into the base of masonry walls through capillary action.

Salt Fretting: A condition of masonry resulting in an obvious pattern of erosion caused by salts, usually from the salting of icy sidewalks.

Spalling: A condition of masonry in which the outer layer or layers of masonry units begin to break off (unevenly) or peel away in parallel layers from the larger block of masonry.

Staining: A discoloration of masonry arising from foreign materials.

Sugaring: A characteristic of some masonry indicative of gradual surface disintegration, sometimes in a powdery condition.

Surface Condensation: An accumulation of moisture or water droplets that occurs on the surface of masonry when moist air is cooled to the point at which the water vapor (a gas) changes state into water (a liquid).

- The kinetics of surface condensation and evaporating is controlled by the surface temperature rather than the ambient temperature (Haynie, 1980).

Surface Crust: An accumulation of the products of efflorescence of the masonry surface to a measurable thickness.

Surface Induration: A hardened deposit of surface crust.

Surface Temperature: The physical temperature of the masonry at its surface as distinguished from air temperature. Continuous monitoring of temperature is usually required in investigation of condensation problems.

Tidemark: A stain on a masonry wall that occurs where equilibrium has been reached between capillarity and evaporation.

Transition Area: The area above the tidemark on a masonry wall that reflects a changing tidemark as equilibrium conditions vary with the weather.

Warpage: The curvature of a flat masonry unit measured as a deviation from a true plane along the edges or the diagonals of the unit.

Water Penetration/Permeance: Physical passage of water through a masonry assemblage by means of interconnected pores (measured by the permeability of the materials) or through cracks and other physical openings.

- Experience indicates that water penetration through masonry exterior walls is the most common cause of complaints regarding nonperformance (Section 4.9.1, Raths, 1985).
- The severity of mechanical damage due to freezing and thawing is directly proportional to the water content of the porous solid.

Weathering: Natural disintegration and erosion of masonry caused by wind and rain, resulting in granular and rounded surfaces.

- Weather as the main general environment normally is defined as the state of the atmosphere with respect to temperature, humidity, precipitation, wind, and radiation at a specific time and place (Garden, 1980).

3.4.3.5 Physical Conditions of Masonry (Structural)

The physical conditions of masonry that are influenced by structural conditions (movement of support, insert slippage, settlement, overloads, volume changes in the structure due to elastic shortening, creep, or thermal changes, etc.) include the following:

Bowing: An outward swelling, or protuberance of a portion of a masonry assemblage.

Bulging: An outward swelling, bowing, or protuberance of a portion of a masonry assemblage from a vertical plane.

Chipping: A condition of small pieces or larger fragments of masonry units separating from the unit resulting from excessive stress.

Condition of Lintels (Shelf Angles): Vertical, horizontal, or rotational movement of the structural members supporting masonry assemblages including the connections and connectors of the supporting members.

- Causes of movement are corrosion and rustjacking.

Cracking: Narrow fissures in masonry 0.0004 in. (0.01 mm) or larger in width.

- Structural behavior affecting masonry walls reflects the integrated behavior of the wall with the supporting structural frame. When the behavior of the supporting frame and masonry wall are not compatible, cracking results (Section 4.9.1; Raths, 1985; Stockbridge, 1980).

Deflection/Settlement: Vertical movement of the structure supporting a masonry assemblage. Differential movement is included as well as total movement.

Detachment: The result of a complete break (or failure of an original construction joint) in a masonry unit in which the detached portion of masonry survives intact.

Displacement: Horizontal translation of all or of a portion of a masonry assemblage.

Distortion: A change in shape of a masonry assemblage. Distortion can include bowing, bulging, sagging, and warping.

Plumbness: Deviation of a surface of a masonry assemblage from the vertical. Masonry can be out of plumb and still exhibit no signs of bowing, bulging, sagging, warping, or other distortions.

Sagging: A vertical movement of a masonry assemblage characterized by wavy horizontal lines in the surface often accompanied by other modes of distortion.

Spalling: A condition of masonry in which the outer layer or layers of masonry units begin to break off (unevenly) or peel away in parallel layers from the larger block of masonry. Often an indication of overstress of the masonry unit.

Ties and Anchors: Units installed in mortar joints to prevent separation of masonry assemblages at control joints, corners, and wall intersections. Units may be metal tie bars, metal lath, metal strips, wire mesh, or hardware cloth.

- Half cell potential values cannot predict corrosion of reinforcement in masonry structures (Haver, et al., 1990).

Warping: The curvature of a masonry assemblage measured as a deviation from a true plane along the edges or the diagonals of the assemblage.

3.4.3.6 Physical Conditions of Masonry (Architectural)

The physical condition of associated architectural features and details that have a significant influence on the physical conditions of masonry include the following:

Condition of Flashing: The degree of the loss of integrity of the thin impervious material placed in construction joints (e.g., in mortar joints and through air spaces in masonry walls) to prevent water penetration and/or provide water drainage.

Condition of Roofing: The degree of the loss of integrity of the material used as roof covering (such as shingles, slate, sheet metal, tile, built-up roofing, etc.) to make it waterproof.

Control Joints: Joints between adjacent parts of masonry which accommodate contraction of adjacent materials thereby avoiding the development of high stresses and cracks.

Damp Course: An impervious horizontal layer of material in a masonry assemblage used to prevent the capillary flow of moisture; usually from the ground but also used to prevent seepage from above.

Dampproofing: A treatment to the surface of a masonry assemblage used to retard the passage or absorption of water or water vapor.

Expansion Joints: Joints or gaps between adjacent parts of masonry which accommodate expansion or contraction of adjacent materials (BIA Technical Note 18a, 1991).

Horizontal Expansion Joints: In masonry wall assemblages, a sealed joint (no mortar or grout) left at specified vertical intervals (usually immediately below shelf angles) to allow for deflection of the structure, to prevent the mass of the masonry above from being transmitted to the masonry below, and to account for relative movement of components. Horizontal expansion joints (also called pressure-relieving joints) are usually filled with a compressible material and sealed at the surface to exclude moisture.

Location: The position of a portion of the masonry assemblage in relation to the entire building structure. The location may have a significant influence on physical conditions.

3.4.3.7 Physical Conditions of Mortar and Grout

Physical conditions of mortar and grout include the following:

Air Content: The volume of air voids in the mortar or grout, usually expressed as a percentage of the total volume.

Alignment: The theoretical, definite lines that establish the position of construction of a masonry assemblage.

Bed Joint Levelness: The levelness or deviation from a horizontal plane of the layer of mortar on which masonry units are set.

Bed Width Variation: The variation in thickness of the layer of mortar on which masonry units are set.

Condition: A general overall assessment of the state of the mortar or grout and its ability to function as intended.

Cracking: Narrow fissures from 0.0004 in. (0.01 mm) or larger in width in the mortar or grout.

Exposure of Joint Reinforcement: Visible exposure of reinforcing steel bars, masonry ties, anchors, and joint reinforcement providing an opportunity for rapid corrosion of the reinforcement.

Friability: Mortar or grout that is easily crumbled or pulverized and easily reduced to powder.

Hardness: A term relating to the capacity of mortar or grout to withstand permanent deformation. Mortar or grout that is too hard, especially differential in hardness along a masonry joint, can lead to an overstressing condition in the masonry units.

Pitting: The development of small cavities in a surface of grout and mortar owing to phenomena such as corrosion, cavitation, or local disintegration.

Sandiness: The presence of sand on the exposed surfaces of mortar and grout indicative of cement deficiency or excessive acid washing by cleaning or rain.

Voids: In mortar or grout, the air spaces between and within pieces of aggregate or between units and mortar.

3.4.4 Test Methods

Visual nondestructive and destructive tests are used to establish properties and physical conditions of masonry structures. The most common test procedures including requirements, advantages, and limitations are listed in Section 3.7, Tables 3.7.8 and 3.7.9.

3.4.5 Combination of Test Methods

It may be necessary to use a combination of methods for satisfactorily predicting properties of masonry (see Section 3.2.4).

3.4.6 References

References cited in Section 3.4, unless otherwise noted, are included in Section 3.6. Additional references are to be found at the end of Chapter 4 and after each of the tables in Section 3.7.

3.5 CONDITION ASSESSMENT OF WOOD

3.5.1 Introduction

Wood is a biological rather than a manufactured material. Wood products are manufactured. Being a biological material, wood possesses inherent defects, such as knots and splits, and a variability in species that must be considered in evaluating engineering property characteristics. Another biological characteristic is that wood is susceptible to attack by various organisms.

Wood is an orthotopic, nonhomogeneous material whose properties are dependent on the fiber orientation in a member. The term "fiber" is used synonymously with "grain," meaning direction corresponding to the long axis of the wood cells. Many species, both softwood and hardwood, are available for construction. Several products, such as solid wood lumber, glued laminated timbers (glulam), laminated veneer lumber (LVL), composite and plywood panel products, and wood trusses are found in building construction.

The condition assessment of wood in buildings requires a determination and evaluation of the load history, identification of the species and grade of wood, and an assessment of the properties and physical condition of the wood and of its connections. The physical condition is influenced, in part, by fire, fungus and insect damage, moisture content, load history, creep, and chemical treatments. Major properties and physical conditions of wood are discussed in Section 3.5.3, Section 3.7, and Tables 3.7.11 and 3.7.12 which provide a guide to selecting test methods which can be used to evaluate properties and the extent of defects.

Basic information necessary for conducting an engineering evaluation of wood structures is contained in the National Design Specification and Design Values for Wood Construction (NFPA, 1991); Evaluation Maintenance and Upgrading Wood Structures (Section 3.7, Table 3.7.12, ASCE Technical Committee on Wood, 1982), Wood Handbook; Wood as an Engineering Material (Section 3.7, Table 3.7.11, U.S. Department of Agriculture, 1987); and Inspection of Wood Beams and Trusses (Section 3.7, Table 3.7.12, NFEC, 1985).

3.5.2 Manufactured Wood Products

Solid Sawn Wood: Lumber and timber for construction are graded under the national grading rules promulgated within the American Softwood Lumber Standard (U.S. Department of Commerce, 1986, Section 3.7, Table 3.7.12). There is a single set of names and descriptions for lumber graded under these rules. Stress grades are assigned to individual members by visual grading procedures (e.g., Select Structural, No. 1, No. 2, etc.) and/or by nondestructive testing equipment (e.g., machine stress rated grades 1650 F, 2100 F, 2400 F, etc.). These stress grades are typically indicated by a grade stamp on the member. Lumber grade names and classifications are given in various grading agency specifications (see References of Section 3.7, Table 3.7.12). Lumber dimensions are given as nominal sizes but actual dimensions are different. Actual dimensions are given in the National Design Specification and Design Values for Wood Construction (NFPA, 1991).

Structural Composite Lumber Including Laminated Products: Structural composite lumber including laminated products is made by bonding layers of wood and/or small wood particles (chips, flakes, etc.) together to form dimensioned lumber. If properly manufactured, such products have reduced material property variability compared to solid sawn wood.

(a) Structural Glued Laminated Timber (Glulam). These members are made by gluing laminations of solid sawn lumber to form the structural member. The grain of the lamination is parallel to the longitudinal axis of the member. Grades and stress classifications of glulam timbers are described by American Institute of Timber Construction (AITC, 1994) specifications and manufacturers product catalogs.

(b) Laminated Veneer Lumber (LVL) is made by gluing thin veneers with the grain of all veneers parallel to the longitudinal axis of the member. Uses for LVL include joists, beams, and truss chords. Grades and classifications for LVL are described by manufacturers product standards.

Trusses: Several types of wood trusses exist. These include:

(a) Heavy timber trusses connected with bolts, split rings, and heavy metal plates.
(b) Light wood trusses connected with toothed metal plates.
(c) Metal-webbed wood-chord trusses.

The design of heavy timber trusses is described in various references (Beyer, 1988; Hoyle and Woeste, 1989; Gusfinkel, 1973; AITC, 1994). Information on manufactured trusses using toothed metal plates or metal webs can be found in individual manufacturers product catalogs or from the Truss Plate Institute (TPI, 1985).

Structural Wood Panel Products:

(a) Plywood. Plywood is made using a balanced construction of thin veneers glued with their grain alternating 90 degrees in adjacent layers. Plywood grades and stress classifications are described by the APA—The Engineered Wood Association (APA, 1994).

(b) Composite Panels. Composite panels are manufactured by bonding veneer faces to reconstituted wood cores or bonding with adhesives small wood particles (chips, flakes, etc.). Examples include oriented strand board, particleboard, and waferboard. Grades and classifications for composite panels are described by the APA—The Engineered Wood Association (APA, 1994).

Combined Wood Structural Elements: Highly efficient structural products are manufactured by combining different wood elements. Examples include structural composite lumber, wood I-beams, built-up beams, and stress-skin panels. Product classifications are described by manufacturers product information. Structural components combining wood and other materials are found in building construction including timber arches, rigid frames tied together with steel rods and flitch beams.

3.5.3 Properties and Physical Conditions of Wood

The determination of the condition of wood in a structure requires the evaluation of the structural history and the assessment of the physical and mechanical properties and the assessment of the physical conditions. Properties and physical conditions are defined and discussed in this Section.

3.5.3.1 Physical Properties of Wood

Classification: Woods are classified into two broad groups, softwoods and hardwoods (Section 3.7, Table 3.7.11, U.S. Department of Agriculture, 1987). The hardwood/softwood terminology has no reference to the actual hardness of the wood. Wood is classified into species for further assignment of physical and mechanical properties.

(a) Softwoods are generally cone-bearing trees with needle-like leaves. The majority of wood structural members are manufactured from softwoods. Examples include Douglas Fir and Southern Pine.

(b) Hardwoods are deciduous and have broad leaves. These woods are generally used architecturally for flooring, trim and paneling. However, they are also used as load-carrying members. Examples include oak, ash, and walnut.

Density: The oven-dry mass of solid wood is termed density, which is used to establish the specific gravity of wood. Many physical and mechanical properties are directly related to density.

Grade: Visual grading is the oldest stress-grading method. It is based on the premise that mechanical properties of lumber differ from mechanical properties of clear wood based upon the many growth characteristics which can be seen and judged visually. These growth characteristics are used to sort lumber into stress grades. To be graded and marked with a grade stamp, all six surfaces of a member must be examined for conformance with explicit grading rules. Only a certified grader may stamp with a grade stamp. A visible grade stamp on a member provides information on which to base an evaluation. Without a grade stamp, a grader or a professional with wood technology experience may be needed to assign a grade. Machine stress rating is based on an observed relation between modulus of elasticity and bending strength of lumber. Modulus of elasticity along with visual grading are the sorting criteria used in this method of grading. Machines operating at high rates of speed measure the modulus of elasticity of individual pieces of lumber, and sort according to a bending strength/stiffness grade class.

Growth Characteristics: Physical and mechanical properties can be influenced by growth characteristics which can include knots and sloping grain. A more complete description of these characteristics can be found in the Wood Handbook (Section 3.7, Table 3.7.11, U.S. Department of Agriculture, 1987).

(a) Heartwood. The inner portion of the tree, consisting of inactive cells that have undergone physical and chemical changes; in many species heartwood is darker than sapwood. Heartwood of most species is more resistant to decay than sapwood.

(b) Knots. Knots are a growth characteristic of wood which may, depending on their size, number, and location in a member, adversely affect the strength properties of the member.

(c) Sapwood. The area between the bark and the heartwood. Usually lighter in color than the heartwood.

(d) Slope of Grain. A growth characteristic indicated as a deviation of the grain from the longitudinal axis of the member. The slope of grain severely affects the physical and mechanical properties of wood.

(e) Juvenile Wood. Wood of lower quality located near the center (pith) of the tree. The presence of juvenile wood can severely affect the physical and mechanical properties of the wood (Section 3.7, Table 3.7.11, U.S. Department of Agriculture, 1987). Other growth characteristics which influence mechanical and physical properties are discussed in depth in several references (Hoyle and Woeste, 1989; Section 3.7, Table 3.7.11, U.S. Department of Agriculture, 1987).

3.5.3.2 Mechanical Properties of Wood

Wood is orthotopic, i.e., it has different properties in the direction of the three mutually perpendicular axes; longitudinal, radial, and tangential. The fundamental mechanical properties of wood are determined from tests performed on small, clean, straight-grain samples (Section 3.7, Table 3.7.11, U.S. Department of Agriculture, 1987). Design values for structural wood elements, which are considerably lower than clear wood properties, are referenced in the NDS or TCM (NFPA, 1991; AITC, 1994).

Bending Strength: Allowable bending strength is based upon a percentage of the ultimate load-carrying capacity as indicated by a modulus of rupture. When three or more wood members (e.g., joists) are spaced not more than 24 in. apart and joined or covered by some type of load-distributing element, the allowable bending stress is 15% higher than for single members to account for load sharing.

Compression Parallel to Grain: A measure of wood's resistance to axial stresses along the axis of the fibers. These stresses are important in evaluating compression members and bearing capacity of the fasteners.

Compression Perpendicular to Grain: A measure of wood's resistance to load when applied perpendicular to the axis of the fibers. An example is a beam resting on a post. The length and location of the bearing area are factors in determining the allowable stress.

Fatigue Strength: When the repetitions of design stress are expected to be more than 100,000 cycles during the normal design life, an allowable design stress less than the allowable static design stress should be considered.

Modulus of Elasticity: The modulus of elasticity of wood is dependent on the orthotopic direction under consideration. However, design moduli of elasticity are based upon bending tests of lumber and are usually only given for the longitudinal direction. Modulus of elasticity for wood is not adjusted for duration of load.

Shear Strength: Horizontal shear is developed from parallel-to-the-grain bending stresses. Because of the structure of wood, shear resistance along the fibers controls shear capacity. Specific concerns include notches and holes in beams.

Tensile Strength Parallel to Grain: Clear wood intrinsically has high tensile strength parallel to the wood fibers. However, growth and manufacturing defects may markedly reduce a member's tensile strength when compared to clear wood strength.

Tensile Strength Perpendicular to Grain: Tension perpendicular to the grain is a measure of the resistance of wood to forces acting across the wood fibers that tend to cause splits in a member. This is typically the weakest mechanical property for wood and, when exceeded, results in brittle fracture. Many structural failures can be traced to excessive tensile stresses perpendicular to the grain. Examples include radial stresses in curved glulam beams; combined shear and tension perpendicular to grain stress at notches and connector holes; and restrained shrinkage across the grain at connections.

3.5.3.3 Properties of Connections

Connections for structural wood are usually made of metal or wood. Common connections include hangers for joists and purlins, framing anchors, metal and wood side plates, bent straps, U-straps, bands, clip angles, bearing plates, steel saddles, hanger rods, arch shoes and steel and wood gusset plates. Steel side plates are quite often used with shear plate connectors. Connections require connectors to hold the several parts together. Design criteria for wood connections are given in detail in the National Design Specification (NFPA, 1991). Reference is made to Section 3.3.2.1 for properties of metal connections and to Sections 3.5.3.1 and 3.5.3.2 for physical and mechanical properties of wood.

3.5.3.4 Properties of Connectors

Connectors are used to make connections. Connectors are also used to fasten wood members together. Connectors include bolts, drift bolts, nails, screws, lag screws, spikes, staples, split ring connectors, toothed ring connectors, metal plate connectors (truss plates) and shear plate connectors.

Chemical Composition: Indicators of the properties of bolts (including washers and nuts), nails, rods, screws, and timber connectors which can be used to establish corrosion resistance and mechanical properties.

Elongation: Elongation, a requirement for certain types of timber connectors, is the increase in gauge length in a tension test specimen measured after fracture as a percentage of the original length. It is a measure of ductility.

Hardness: For threaded fasteners sufficient hardness insures the surfaces of the fasteners have the capacity to transmit load by bearing; it is a specified requirement for nuts and washers and sometimes for bolts and rods.

Lateral Resistance: The resistance to lateral movement of nails, screws, or bolts which are embedded in wood. Connections in wood develop lateral resistance by fastener bearing. The resistance is dependent upon the species (density), fastener diameter and yield strength, and the location of the fastener with respect to the edges of connected members, the depth of fastener penetration in the wood, and orientation of the fastener with respect to the grain of the wood.

Proof Load: A specified load without measured permanent set; applicable to bolts, rods and nuts.

Resistance to Delamination: A measure of the integrity of glue joints for structural laminated members expressed as a percentage of total length of open joint after the joint is subjected to an autoclave test in water at specified temperature.

Tensile Strength: Stress calculated from the maximum load sustained by a metal connector during a tension test divided by the original cross-sectional area of the specimen; a required physical property for bolts, rods, and some types of timber connectors. For certain materials an equivalent tensile strength is determined.

Tensile Strength of Fastener: Maximum stress calculated from the maximum load sustained by a metal fastener during a tension test divided by the original cross-sectional area of the fastener. Tensile strength may limit the fastener's ability to develop the strength of the wood when used as a fastener.

Timber Joint Compressive Strength: A full-scale test of a timber joint to determine the maximum compressive load the joint can sustain.

Timber Joint Tensile Test: A full-scale test of a timber joint to determine the maximum tensile load the joint can sustain.

Torsional Moment Capacity Test: A full-scale test of a joist or purlin hanger to determine the torsional movement the hanger can sustain.

Vertical Load Capacity Test: A full scale test of a joist or purlin hanger to determine the vertical load the hanger can sustain.

Withdrawal Strength: The resistance to withdrawal of nails and screws in wood. The resistance is dependent upon the length and diameter of the fastener, species and other properties of wood and the location of the nail or screw with respect to the grain of the wood.

Yield Strength: Stress at which a significant increase in strain occurs without a corresponding increase in stress; a property which is important in the evaluation of certain types of timber connectors. For certain materials an equivalent yield strength is determined.

3.5.3.5 Physical Conditions of Wood

Adhesives: Structural elements can be joined together by adhesives rated for exterior or interior use. These adhesives can be affected by moisture content, temperature, and service conditions that may result in their deterioration. Adhesive failure may result in loss of composite action of the interconnected elements.

Chemical Exposure: In general, wood is highly resistant to many chemicals, and heartwood is more resistant to chemical attack than sapwood. Some chemicals can have a deteriorating influence on the wood and result in a change in the mechanical properties with time (Section 3.7, Table 3.7.11, U. S. Department of Agriculture, 1987).

Chemical Treatments: Wood may be chemically treated to enhance its preservation or its flame-retardant qualities. These treatments can change the mechanical properties of the wood and, if highly acidic, can promote the corrosion of connections and connectors, in addition to attacking the wood (NFPA, 1991). A common preservation treatment is chromated copper arsenate (CCA) which gives the wood a green tint. Detailed information about preservative-treated wood is available from the American Wood Preservers Association (AWPA, 1992). Detailed information about fire-retardant treated wood is available from treatment manufacturers.

Creep: Time-dependent deformation under sustained loads. The magnitude is affected by moisture content, temperature, and other factors.

Cross-Sectional Properties: Actual dimensions and other geometric properties of structural members.

Decay: Decay fungi are wood-destroying organisms that use the organic material of wood as food. For growth, they need suitable temperature, sufficient oxygen, and adequate moisture. Wood-moisture content over 20%, temperatures between 40°F (4°C) and 105°F (40°C) and air (oxygen) are the environmental conditions normally needed to support decay. Remove one of these conditions and wood will not decay. Wood protected from decay and other deteriorating influences shows little change in mechanical properties with time. The detection and extent of decayed wood is critical to condition assessment because wood decay seriously reduces all mechanical properties.

Deflection: Measured deflection usually reflects both elastic and creep deformation.

Duration of Load: Load duration is the length of time a member supports a load. The load required to produce failure over a long period of time is less than the load required to produce failure over a shorter period. Allowable design stresses are adjusted to the assumed duration of load used in design (NFPA, 1991).

Insect and Animal Attack: Insects and animals can attack and destroy wood. Examples are termites, marine borers, carpenter ants, and rodents. Loss of cross-section area can result from insect and animal attack.

Moisture and Its Effects: Moisture has a dramatic effect on properties of wood. The moisture content of wood is the mass of water in the wood expressed as a percentage of the oven-dry mass of the wood.

(a) Pattern of Exposure to Moisture. If wood changes moisture content after installation, the resulting shrinking and swelling can cause distortion and twisting. This change in dimension only occurs at moisture contents below fiber saturation. Fiber saturation is the stage in the drying or wetting of wood at which the cell walls are saturated and the cell cavities are free from water. It is usually taken as approximately 30% moisture content. A decrease in moisture content below the fiber-saturation point will increase the strength of wood and will cause it to shrink. A change in moisture content above the fiber-saturation point will not influence the strength or the shrinkage.

(b) Effects of Changing Moisture Content. The major effects of changing moisture content of wood are:

(1) *Warp*. Any variation from a plane surface including bow, crook, cup, or twist as a result of (but not limited to) a change in moisture content. Warp can result in eccentricity of applied load. Deformations due to bowing and crooking must be considered when evaluating beam, column, and pile strengths.

(2) *Checks, Splits*. Checks and splits can result from differential shrinkage of wood in the radial and tangential directions of the growth rings. Splits and checks are wood failure, and their effect should be considered in the assessment process.

Physical Damage: Wood can be physically damaged due to fire, impact, abrasion, deliberate alteration or other attacks on the original cross-sectional properties. This physical damage can adversely affect the performance of the wood member.

Temperature Environment: The mechanical properties of wood are reduced under high and sustained temperatures.

Weathering: Exposed wood will turn gray when exposed to the weather. In the weathering process, some wood will lose only 0.25 in. (6 mm) of thickness in 100 years. Wood not well suited to resist weathering may deteriorate markedly in a short period of time.

3.5.3.6 Physical Conditions of Connections

Connections for structural wood are usually made of metal or of wood. Reference is made to Section 3.3.2.4 for the physical condition of metal connections and to Section 3.5.3.5 for the physical condition of wood connections.

3.5.3.7 Physical Conditions of Connectors

Angle to Grain Load: The angle formed between the axis line of the connector and the longitudinal axis of the wood fibers.

Condition of Connector: The physical state as determined by observation and with use of simple physical techniques such as cleaning, scraping, and sounding; can be used to establish existence of many of the physical conditions included herein.

Condition of Glue: The condition of the glue in laminated or built-up members measured by the percentage of the open joint surface or the total surface of joints that should be glued.

Corrosion: Same as electrolytic or electrochemical corrosion in physical condition of metals (Section 3.3.2.4). Both the metal fastener and the wood in the vicinity of the fastener hole can be degraded by metal corrosion.

Cross-Sectional Properties: Actual dimensions and other geometric properties of connectors.

Deformation: Elongation, bending or twisting of bolts, rods, and other connectors over the elastic limit.

Detailing: Actual diameter and length of bolts, rods, timber connectors, including edge distance, spacing, pitch and end distances of bolts, nails, screws, and timber connectors. Minimum spacing and end-distance requirements are critical for developing the connector design capacity.

Eccentricity: The difference between the center of gravity of the group of connectors used in a connection and the line of action of the load on the connection. If more than one member meets at a connection then eccentricity of the group of connectors with each member is possible. Eccentricity must be evaluated in three dimensions. Results of eccentricity include tension perpendicular to grain stresses which can cause members to split.

Elongated Bolt Holes: The condition of oversized bolt holes that develops from insufficient bearing capacity of the wood due to applied loads, or from incorrectly drilled holes. This can cause unequal load distribution among bolts in a row.

Penetration of Nails and Screws: The actual depth of physical penetration of the connector into the wood.

Splits at Connection: The existence of splits along the fibers at the connections is usually caused by either external loads or wood shrinkage.

Tightness: The physical condition of bolts, nails, rods, screws, and timber connectors that indicate the connection fits snugly. The condition of the entire connection is determined by the tightness of the connectors.

3.5.4 Test Methods

Visual, nondestructive, and destructive tests are used to establish properties and physical conditions of wood structures. Most common test procedures including requirements, advantages, and limitations are listed in Section 3.7, Tables 3.7.11 and 3.7.12.

3.5.5 Combination of Test Methods

It may be necessary to use a combination of methods for satisfactorily predicting properties of wood (see Section 3.2.4).

3.5.6 References

References cited in Section 3.5, unless otherwise noted are included in Section 3.6. Additional references are to be found at the end of Section 4 and after each of the tables in Section 3.7.

3.6 REFERENCES

ACI 201-92. (1992). "Guide to Durable Concrete," American Concrete Institute, Farmington Hills, Michigan.

ACI 201.1R-92. (1992). "Guide for Making Condition Survey of Concrete in Service," American Concrete Institute, Farmington Hills, Michigan.

ACI 318-95. (1995). "Building Code Requirements for Reinforced Concrete and Commentary," American Concrete Institute, Farmington Hills, Michigan.

ACI 515.1R-93. (1993). "Protective Systems for Concrete," ACI Committee Report, American Concrete Institute, Farmington Hills, Michigan.

ACI 530/ASCE 5/TMS 402. (1995). "Building Code Requirements for Masonry Structures," American Society of Civil Engineers, Reston, Virginia.

ACI 530.1/ASCE 6/TMS 602. (1995). "Specification for Masonry Structures," American Society of Civil Engineers, Reston, Virginia.

ACI Monograph No. 6. (1971). "Hardened Concrete: Physical and Mechanical Aspects," American Concrete Institute, Farmington Hills, Michigan, 260 pp.

AISC-LRFD. (1994). "Manual of Steel Construction," American Institute of Steel Construction, Inc., Chicago, Illinois.

AISC-ASD. (1989). "Manual of Steel Construction," American Institute of Steel Construction, Chicago, Illinois.

AITC. (1994). "Timber Construction Manual," American Institute of Timber Construction, John Wiley & Sons, Inc., New York, New York.

APA. (1994). "Design Construction Guide, Residential and Commercial," APA—The Engineered Wood Association, Tacoma, Washington.

ASCE 7-95. (1995). "Minimum Design Loads for Buildings and Other Structures," American Society of Civil Engineers, Reston, Virginia.

ASTM C270. (1997). "Standard Specification for Mortar for Masonry," 1997 Annual Book of ASTM Standards, *Vol. 4.05*, American Society for Testing and Materials, West Conshohocken, Pennsylvania, 136–146.

ASTM C476. (1995). "Standard Specification for Grout for Masonry," 1997 Annual Book of ASTM Standards, *Vol. 4.05*, American Society for Testing and Materials, West Conshohocken, Pennsylvania, 274–275.

ASTM C567. (1991). "Test Method for Unit Weight of Structural Lightweight Concrete," 1997 Annual Book of ASTM Standards, *Vol. 4.02*, American Society for Testing and Materials, West Conshohocken, Pennsylvania, 280–282.

ASTM C1142. (1995). "Standard Specification for Extended Life Mortar for Unit Masonry," 1997 Annual Book of ASTM Standards, *Vol. 4.05*, American Society for Testing and Materials, West Conshohocken, Pennsylvania, 730–733.

AWPA. (1992). "Structural Lumber Fire-Retardant Treatment by Pressure Processes," American Wood Preserver's Association, *Standard C20*, "Plywood, Fire Retardant Treatment by Pressure Processes," Washington, D.C.

Bailey, W. G., Matthys, J. H., and Edwards, J. E. (1990). "Initial Rate of Absorption of Clay Brick

Considering Both Bed Surfaces in the as Received Condition and After Outside Exposure," Masonry; Components to Assemblages, edited by John H. Matthys, *ASTM STP 1063*, American Society for Testing and Materials, West Conshohocken, Pennsylvania, 5–26.

Beyer, D. E. (1988). "Design of Wood Structures," McGraw-Hill Publishing Company, New York, New York.

BIA Technical Note 18a. (1991). "Movement—Design and Detailing of Movement Joints," Brick Institute of America, Reston, Virginia.

BOCA. (1993). "BOCA National Building Code," Building Officials and Code Administrators International, Inc., Country Club Hills, Illinois.

Cady, P. D., and Weyers, R. E. (1983). "Chloride Penetration and the Deterioration of Concrete Bridge Decks," ASTM, Vol. 5, No. 5, American Society for Testing and Materials, West Conshohocken, Pennsylvania, 81–87.

Carrier, R. F., Pu, D. C., and Cady, P. D. (1975). "Moisture Distribution in Concrete Bridge Decks and Pavements," Durability of Concrete, ACI *Publication SP-47*, American Concrete Institute, Farmington Hills, Michigan, 169–190.

Evans, R. M. (1993). "Polyurethane Sealants—Technology and Applications," Technomic Publishing Company, Inc., Lancaster, Pennsylvania, 186 pp.

Evans, R. H., and Marathe, M. S. (1968). "Microcracking and Stress-Strain Curve for Concrete in Tension," Materials and Structures, Research and Testing, RILEM (Paris) Vol. 1, No. 1, January–February, 61–64.

EWS. (1995). Glued Laminated Beam Design Tables for Western Species, American Wood Systems, Form EWS 5475, Tacoma, Washington.

FEMA 178. (1992). "NEHRP Handbook for the Seismic Evaluation of Existing Buildings," Building Seismic Safety Council, Washington, D.C.

Garden, G. K. (1980). "Design Determines Durability," Durabilities of Building Materials and Components, ASTM *STP 691*, edited by P. J. Sereda and G. G. Litvan, American Society for Testing and Materials, West Conshohocken, Pennsylvania, 31–37.

Gerwick, B. C., Jr. (1991). "The Global Advance—Emerging Opportunities of Home and Abroad," *PCI Journal*, Vol. 36, No. 6, Precast/Prestressed Concrete Institute, Chicago, Illinois, November–December, 32–37.

Gopalaratnam, V. S., and Shah, S. P. (1985). "Softening Response of Plain Concrete in Direct Tension," *ACI Journal*, Title No. 82-27, American Concrete Institute, Farmington Hills, Michigan, May–June, 310–323.

Gusfinkel, G. (1973). "Wood Engineering," Southern Forest Products Association, New Orleans, Louisiana.

Hamid, A. A., and Drysdale, R. G. (1980). "Effect of Strain Gradient on Tensile Strength of Concrete Blocks," Masonry: Materials, Properties, and Performance, *ASTM STP 778*, edited by J. G. Borchelt, American Society for Testing and Materials, West Conshohocken, Pennsylvania, 57–65.

Hatzinikolas, M., Longworth, J., and Warwaruk, J. (1980). "Properties of Ready-Mixed Mortar," Masonry: Materials, Properties, and Performance, *ASTM STP 778*, edited by J. G. Borchelt, American Society for Testing and Materials, West Conshohocken, Pennsylvania, 15–26.

Haver, C. A., Keeling, D. W., Somayaji, S., Jones, D., and Heidersbach, R. H. (1990). "Corrosion of Reinforcing Steel and Wall Ties in Masonry Systems," Masonry: Components to Assemblages, *ASTM STP 1063*, edited by J. H. Mathys, American Society for Testing and Materials, West Conshohocken, Pennsylvania, 173–193.

Haynie, F. H. (1980). "Theoretical Air Pollution and Climate Effects on Materials Confirmed by Zinc Corrosion Data," Durabilities of Building Materials and Components, ASTM *STP 691*, edited by P. J. Sereda and G. G. Litvan, American Society for Testing and Materials, West Conshohocken, Pennsylvania, 157–175.

Hoyle, R. J., and Woeste, F. E., Jr. (1989). "Wood Technology in the Design of Wood Structures," Fifth Edition, Iowa State University, Ames, Iowa.

Hughes, B. P., and Chapman, G. P. (1966). "The Complete Stress-Strain Curve for Concrete in Direct Tension," *RILEM Bulletin (Paris) New Series No. 30*, March, 95–97.

Isberner, A. W., Jr. (1980). "Grout for Reinforced Masonry," Masonry: Materials, Properties, and Performance, *ASTM STP 778*, edited by J. G. Borchelt, American Society for Testing and Materials, West Conshohocken, Pennsylvania, 38–56.

Isberner, A. W. (1983). "A Test Method for Measuring the Efflorescence Potential of Masonry Mortars," Masonry: Research, Application, and Problems, *ASTM STP 871*, edited by J. C. Grogan and J. T. Conway, American Society for Testing and Materials, West Conshohocken, Pennsylvania, 27–37.

Litvan, G. G. (1980). "Freeze-Thaw Durability of Porous Building Materials," Durabilities of Building Materials and Components, ASTM *STP 691*, edited by P. J. Sereda and G. G. Litvan, American Society for Testing and Materials, West Conshohocken, Pennsylvania, 455–463.

Mather, B. (1968). "Cracking Induced by Environmental Effects," Causes, Mechanism, and Control of Cracking in Concrete, American Concrete Institute, Farmington Hills, Michigan, 67–73.

Mather, B. (1989). "Concrete Strength-Quality Assurance," Structural Materials, *Proc., Sessions Related to Structural Materials at Structures Congress '89*, edited by J. F. Orofino, American Society of Civil Engineers, Reston, Virginia, 442–450.

NFPA. (1991). "National Design Specification and Design Values for Wood Construction," National Forest Products Association, Washington, D.C.

NPA. (1982). "Standard for Particle Board Decking for Factory-Building Housing," National Particleboard Association, Silver Spring, Maryland.

Plowman, J. M. (1956). "Maturity and the Strength of Concrete," Magazine of Concrete Research, Vol. 8, No. 22, March, 13–22.

SBCC. (1994). "Standard Building Code," Southern Building Code Congress International, Birmingham, Alabama.

Shah, S. P., and Winter, G. (1968). "Inelastic Behavior and Fracture of Concrete," Causes, Mechanism, and Control of Cracking in Concrete, ACI, *Publication SP-20*, American Concrete Institute, Farmington Hills, Michigan, 5–28.

SHRP-P-338 (FHWA-FA-94-086). (1993). "Distress Identification Manual for the Long-Term Pavement Project," Strategic Highway Research Program, National Research Council, Washington, D.C., 147 pp.

Stockbridge, J. G. (1980). "Evaluation of Terra Cotta In-Service Structures," Durabilities of Building Materials and Components, *ASTM STP 691*, edited by P. J. Sereda and G. G. Litvan, American Society for Testing and Materials, West Conshohocken, Pennsylvania, 218–230.

Stutzman, P. E., and Clifton, J. R. (1994). "Diagnosis of Causes of Concrete Deterioration in the MLP-7A Parking Garage," NISTIR 5492, NIST, Gaithersburg, Maryland.

Suprenant, B. A., and Schuller, M. P. (1994). "Nondestructive Evaluation and Testing of Masonry Structures," The Aberdeen Group, Addison, Illinois, 194 pp.

TPI. (1985). "Design Specifications for Metal Plate Connected Wood Trusses," Truss Plate Institute, Madison, Wisconsin.

UBC. (1994). "Uniform Building Code," International Conference of Building Officials, Whittier, California.

Weyers, R. E., and Smith, D. G. (1989). "Chloride Diffusion Constant for Concrete," Structural Materials, *Proc., Sessions Related to Structural Materials at Structures Congress '89*, edited by J. F. Orofino, American Society of Civil Engineers, Reston, Virginia, 106–115.

Winter, G. (1961). "Properties of Steel and Concrete and the Behavior of Structures," *Transactions*, ASCE, Vol. 126, Part II, Paper No. 3264, American Society of Civil Engineers, Reston, Virginia, 1054–1100.

Wong, H. E., and Drysdale, R. G. (1983). "Compression Characteristics of Concrete Block Masonry Prisms," Masonry: Research, Application, and Problems, *ASTM STP 871*, edited by J. C. Groyan and J. T. Conway, American Society for Testing and Materials, West Conshohocken, Pennsylvania, 167–177.

3.7 TABULATION OF TEST METHODS

Testing of specimens from members of a structure to determine chemical and physical properties can provide useful quantitative data. While removal of specimens may be destructive, careful selection and handling can ensure that the usefulness of the member is not impaired and the damaged areas can be satisfactorily repaired. Tests on specimens provide specific, quantitative information without the requirement of questionable analyses and interpretation of results. Wherever and whenever possible, precedence should be given to physical testing of specimens over nondestructive testing methods to determine similar information. It is also advantageous to use the results of tests in specimens from members as base data to correlate nondestructive testing programs.

Nondestructive methods, while useful in determination of physical properties of members, are generally used to establish physical conditions and extent of deterioration of structural members and systems. Depending upon equipment, results for determination of elastic properties are usually more accurate than those for determining other physical properties. Most nondestructive methods require highly qualified and skilled personnel, specialized equipment and considerable analysis and interpretation of data to develop qualitative and quantitative information.

Available test methods are cited in the following series of tables. It is not intended to provide the reader with information and details on how to perform and interpret tests, but rather to provide information on test methods that are available, what they can accomplish, advantages and limitations of the tests and a brief statement about qualifications of test personnel. This information is subjective, not exhaustive and constantly subject to change. A method that is expensive at this time may be reasonably priced a few years later. Advantages and limitations also change with time. Accordingly, the references are the most important part of these tables and care has been taken to provide complete reference information. Appropriate references are thus provided for in-depth information and perusal of any specific reference will usually lead to an extensive bibliography on the subject. In addition, the supplemental references in Section 3.8 contain further information that can be pursued for in-depth knowledge. The reader is advised not to focus on a specific test method for a specific material but to also review the information given for the same method under other materials when available. Load tests on buildings have been included as part of the subject of test methods.

Each Table in this Section (Tables 3.7.1–3.7.13) is followed by a list of related references. These references are numbered to facilitate referral in the "Reference" row of the Tables of Section 4 as well as to the reference column in the tables in this Section themselves.

This page intentionally left blank

TABLE 3.7.1. Test Methods for Determining Chemical and Physical Properties of Concrete

Property	Purpose of Test	Test Method	User Expertise	Advantages	Limitations	References
Air-void content	Determination of the proportional volume of air-voids in concrete expressed as volume percent of the hardened concrete, includes both "entrapped" and "entrained" air.	Measurement of the air-void system in hardened concrete by prescribed microscopical procedures on sawed and ground sections from specimens taken from the hardened concrete.	Qualified petrographer required.	Common test providing useful information.	Difficulties may arise in preparing a satisfactory surface from the specimen for analysis.	4, 11, 17, 18, 22
Chloride content	Determination of chloride content.	Samples are pulverized, weighed, mixed with water, treated chemically, heated, filtered and volume passed tested for chloride ion content.	Standard laboratory test.	Common test providing useful information.	Many samples required to determine chloride ion profile in concrete.	1
Compressive strength	Determination of compressive strength from drilled cores and portions of beams broken in flexure.	Prepared specimens are tested to fracture in a testing machine.	Standard laboratory test.	Common test providing useful information; low in cost.	Correction factors for specimens with L/D less than 1.8 are approximate and subject to variability.	2, 3, 4, 6, 14, 19, 21, 22
Flexure strength	Determination of flexural strength (Modulus of Rupture) and, indirectly, tensile strength of specimen.	Loading of a standard beam test specimen until fracture occurs.	Standard laboratory test.	Provides useful information.	Precision of test methods have not been established. Not a common test because of size of required specimen.	2, 4, 5, 9, 22
Fundamental frequencies	Determination of significant change in the dynamic Young's modulus of elasticity, dynamic modulus of rigidity, and dynamic Poisson's ratio.	Specimens are forced to vibrate at varying frequencies. Well defined peak of response is fundamental frequency.	Specialized testing ability required.	Assessment of uniformity of concrete.	Costly. Results cannot be used for design.	4, 8

Thickness	Determination of thickness of concrete slabs and walls for compliance with requirements.	Measurement of drilled core length by standard procedures.	Standard laboratory test.	Common test low in cost.	Problem with cores taken from concrete placed against earth due to great number of projections and voids.	4, 7
Petrographic analysis	Determination of acidity, alkali-carbonate reaction; alkali-silica reaction chemical composition; contaminated mixing water and aggregates; quality, uniformity, and soundness of aggregates; water-cement ratio and frozen aggregates.	Prescribed microscopic examination on sawn and ground sections from specimens taken from hardened concrete.	Qualified petrographer required.	Common tests providing much useful information.	Difficulties may arise preparing a satisfactory surface from the specimen for analysis. Costly. Results greatly enhanced if results can be compared to results of other specimens from structure.	4, 11, 17, 18, 20, 21, 22
Potential for volumetric change	Measurement of length change permits assessment of the potential for volumetric expansion or contraction due to causes other than applied force or temperature.	Determination of length changes after both air and water storage.	Required expertise in preparing specimens and setting gauge studs.	Provides useful information.	Costly.	4, 10
Portland cement content	Determination of portland cement content in percentage by mass.	Crushed and graded samples are heated and cooled, subjected to a soluble silica subprocedure, or a calcium oxide procedure, filtration and measurement of silica content of concrete.	Substantial degree of skill required.	Provides very useful information.	Costly. Precision is dependent on composition of concrete and in subprocedure used.	4, 17, 20, 22

SEI/ASCE 11-99

TABLE 3.7.1. Test Methods for Determining Chemical and Physical Properties of Concrete (*Continued*)

Property	Purpose of Test	Test Method	User Expertise	Advantages	Limitations	References
Resistance to freezing and thawing	Determination of the resistance of concrete specimens to rapidly repeated cycles of freezing and thawing.	Test specimens under controlled moisture conditions are subjected to standard cycles of freezing and thawing in water or by rapid freezing in air and thawing in water. Specimens are measured for length and mass before and after tests and percent change recorded as well as a calculated durability factor.	Standard laboratory tests.	Provides useful information on durability and service life.	Precision of results is dependent upon number of tests.	4, 8, 10, 16, 17, 21, 22
Specific gravity (density), percent absorption and percent voids	Determination of specific gravity, percent absorption and voids in hardened concrete.	Specimens are oven dried, mass determined, immersed in water for 48 h, surfaced dried and surface dried mass determined. Specimens are then boiled, surface dry mass recorded, then immersed mass recorded.	Standard laboratory test.	Provides useful information.	Calculation assumes void space in specimen.	4, 15, 21, 22
Splitting tensile strength	Determination of splitting tensile strength which provides measure of tensile strength of hardened concrete specimens.	Standard prepared and cured specimens are tested in a device applying eccentric load.	Standard laboratory test but moderate expertise required.	Provides useful information on tensile strength.	Accuracy required in positioning specimen and use of aligning jig in testing machine for consistent results.	4, 13, 19, 21, 22
Static modulus of elasticity and Poisson's ratio	Determination of chord (Young's) modulus of elasticity and Poisson's ratio.	Standard prepared and cured specimens are loaded in compression and longitudinal and transverse strains measured.	Standard laboratory test.	Provides useful information.	Length to diameter ratio of 1.5 or greater is required for good results.	4, 12, 21, 22

References for Table 3.7.1
1. AASHTO T260. (1994). "Standard Method of Sampling and Testing for Total Chloride Ion in Concrete and Concrete Raw Materials," Standard Specification for Transportation Materials and Methods of Sampling and Testing, Part II, The American Association of State Highway and Transportation Officials, Washington, D.C., 1995 Edition. 650–664.
2. ACI 214.3R. (1988). "Recommended Practice for Evaluation of Strength Test Results of Concrete," ACI Committee 214 Report, American Concrete Institute, Farmington Hills. Michigan.
3. ASTM C39. (1996). "Standard Test Method for Compressive Strength of Cylindrical Concrete Specimens," 1997 Annual Book of ASTM Standards, Vol. 04.02, American Society for Testing and Materials, West Conshohocken, Pennsylvania, 17–21.
4. ASTM C42. (1994). "Standard Test Method for Obtaining and Testing Drilled Cores and Sawed Beams of Concrete," 1997 Annual Book of ASTM Standards, Vol. 04.02, American Society for Testing and Materials, West Conshohocken, Pennsylvania, 24–27.
5. ASTM C78. (1994). "Standard Test Method for Flexural Strength of Concrete (Using Simple Beam with Third-Point Loading)," 1997 Annual Book of ASTM Standards, Vol. 04.02, American Society for Testing and Materials, West Conshohocken, Pennsylvania, 31–33.
6. ASTM C116. (1990). "Standard Test Method for Compressive Strength of Concrete Using Portions of Beams Broken in Flexure." 1997 Annual Book of ASTM Standards, Vol. 04.02, American Society for Testing and Materials, West Conshohocken, Pennsylvania, 52–54.
7. ASTM C174. (1997). "Standard Test Method for Measuring Length of Drilled Concrete Cores," 1997 Annual Book of ASTM Standards, Vol. 04.02, American Society for Testing and Materials, West Conshohocken, Pennsylvania, 109–110.
8. ASTM C215. (1991). "Standard Test Method for Fundamental Transverse, Longitudinal and Torsional Frequencies of Concrete Specimens," 1997 Annual Book of ASTM Standards, Vol. 04.02, American Society for Testing and Materials, West Conshohocken, Pennsylvania, 119–124.
9. ASTM C293. (1994). "Standard Test Method for Flexural Strength of Concrete (Using Simple Beam with Center-Point Loading)," 1997 Annual Book of ASTM Standards, Vol. 04.02, American Society for Testing and Materials, West Conshohocken, Pennsylvania, 163–165.
10. ASTM C341. (1996). "Standard Test Method for Length Change of Drilled or Sawed Specimens of Hydraulic-Cement Mortar and Concrete," 1997 Annual Book of ASTM Standards, Vol. 04.02, American Society for Testing and Materials, West Conshohocken, Pennsylvania, 201–204.
11. ASTM C457. (1990). "Standard Practice for Microscopical Determination of Air-Void Content and Parameters of the Air-Void System in Hardened Concrete," 1997 Annual Book of ASTM Standards, Vol. 04.02, American Society for Testing and Materials, West Conshohocken, Pennsylvania, 225–237.
12. ASTM C469. (1994). "Standard Test Method for Static Modulus of Elasticity and Poisson's Ratio of Concrete in Compression," 1997 Annual Book of ASTM Standards, Vol. 04.02, American Society for Testing and Materials, West Conshohocken, Pennsylvania, 238–241.
13. ASTM C496. (1996). "Standard Test Method for Splitting Tensile Strength of Cylindrical Concrete Specimens," 1997 Annual Book of ASTM Standards, Vol. 04.02, American Society for Testing and Materials, West Conshohocken, Pennsylvania, 263–266.
14. ASTM C617. (1994). "Standard Practice for Capping Cylindrical Concrete Specimens," 1997 Annual Book of ASTM Standards, American Society for Testing and Materials, West Conshohocken, Pennsylvania, 290–293.
15. ASTM C642. (1997). "Standard Test Method for Density, Absorption, and Voids in Hardened Concrete," 1997 Annual Book of ASTM Standards, Vol. 04.02, American Society for Testing and Materials, West Conshohocken, Pennsylvania, 308–309.
16. ASTM C666. (1992). "Standard Test Method for Resistance of Concrete to Rapid Freezing and Thawing," 1997 Annual Book of ASTM Standards, Vol. 04.02, American Society for Testing and Materials, West Conshohocken, Pennsylvania, 310–315.
17. ASTM C823. (1995). "Standard Practice for Examination and Sampling of Hardened Concrete in Constructions," 1997 Annual Book of ASTM Standards, Vol. 04.02, American Society for Testing and Materials, West Conshohocken, Pennsylvania, 396–401.
18. ASTM C856. (1995). "Standard Practice for Petrographic Examination of Hardened Concrete," 1997 Annual Book of ASTM Standards, Vol. 04.02, American Society for Testing and Materials, West Conshohocken, Pennsylvania, 406–422.
19. ASTM C1074. (1993). "Standard Practice for Estimating Concrete Strength by the Maturity Method," 1997 Annual Book of ASTM Standards, Vol. 04.02, American Society for Testing and Materials, West Conshohocken, Pennsylvania, 530–536.
20. ASTM C1084. (1997). "Standard Test Method for Portland-Cement Content of Hardened Hydraulic-Cement Concrete," 1997 Annual Book of ASTM Standards, Vol. 04.02, American Society for Testing and Materials, West Conshohocken, Pennsylvania, 557–562.
21. Bocca, P. (1988). "The Use of Microcores in Structural Assessment," IABSE, Congress Report, 13th Congress, Helsinki, June 6–10, 379–384.
22. Neville, A. M. (1971). "Hardened Concrete: Physical and Mechanical Aspects," ACI Monograph No. 6, American Concrete Institute, Farmington Hills, Michigan, and the Iowa State University Press, Ames, Iowa.

TABLE 3.7.2. NDT Methods for Determining Chemical and Physical Properties of Concrete

Test Method	Purpose of Test	Principle of Operation	User Expertise	Advantages	Limitations	References
Break-off method	Determination of flexural strength.	Break-off of a cylindrical specimen of in-place concrete. Cores are drilled into concrete and broken off with commercial tester.	No specialized expertise required.	Safe, simple and fast to perform requiring only exposed surface.	Limited by maximum aggregate size and minimum member thickness.	8, 14, 23
Nuclear gamma radiation	Determination of in-place density.	A Gamma source or a detector is housed in a probe inserted into a preformed hole in the concrete to a pre-determined depth. Readings are related to density by a calibration curve.	Licensed trained operators are required but minimum operator skills required.	Useful information is obtained and very accurate if calibration curves are established for each project.	Results can be altered by reinforcing steel, chemical composition of concrete constituents and sample heterogeneity.	4, 20
Neutron probe	Determination of chloride content in concrete.	Detection and counting of captured activated gamma rays emitted by chloride ions.	Licensed, trained operators are required but minimum operator skills required.	Can detect very small concentration of chloride content.	In order to avoid modeling the neutron flux distribution, an assumption must be made on chloride ion profile in concrete. The alternate involves considerable computation.	13, 20

Rebound hammer	Compares quality of concrete from different areas of specimen; estimates of concrete strength based on calibration curves with limited accuracy.	Spring driven mass strikes surface of concrete and rebound distance is given in R-values; surface hardness is measured and strength estimated from calibration curves provided by hammer manufacturer.	Simple to operate; can be readily operated by field personnel.	Equipment is lightweight, simple to operate, and inexpensive; large amount of data can be quickly obtained; good for determining uniformity of concrete and areas of potentially low strength.	Results affected by condition of concrete surface; does not give precise prediction of strength; estimates of strength should be used with great care; frequent calibration of equipment required.	3, 7, 8, 11, 12, 15, 16, 19, 21, 26
Resonant frequency testing	Used in the laboratory to determine various fundamental modes of vibration for calculating dynamic module; used in field to detect voids, delaminations, foreign objects, deterioration, and determine thickness.	A resonant frequency condition is set up between two reflecting interfaces. Energy can be introduced by hammer impact or oscillator-amplifier-electromagnetic driver system.	High level of expertise required to interpret results. Technician can be easily trained for laboratory measurements as specimens have simple geometry. Measurements in field require considerable experience.	Allows one to "see inside" concrete structures; can penetrate to depths of a number of feet; a newly developed transducer receiver can improve results over an accelerometer or ultrasonic transducer.	Operates in sonic range and does not have resolution of ultrasonics but numerous reverberations between reflecting surfaces can improve averages. Still in developing stage.	18, 22
Ultrasonic pulse velocity	Gives indication of strength, uniformity and quality of concrete; internal discontinuities can be located and their size estimated; most widely used stress wave method for field use.	Operates on principle that stress wave propagation velocity is affected by quality of concrete; pulse waves are introduced in materials and the time of arrival measured at the receiving surface with a receiver.	Varying level of expertise required to interpret results. Operator requires a fair degree of training.	Equipment relatively inexpensive and easy to operate; accurate assessment of uniformity and quality. By correlating compressive strength of cores and wave velocity, in situ strength can be estimated.	Good coupling between transducer and test substrate critical; interpretation of results can be difficult; density and amount of aggregate moisture variations, and presence of metal reinforcement may affect results; calibration standard required.	1, 8, 9, 10, 12, 15, 19, 24, 25, 26

TABLE 3.7.2. NDT Methods for Determining Chemical and Physical Properties of Concrete *(Continued)*

Test Method	Purpose of Test	Principle of Operation	User Expertise	Advantages	Limitations	References
Ultrasonic pulse-echo	Gives indication of uniformity and quality of concrete. Can give location of rebars; voids in concrete; density and thickness.	Operates on principle that original direction, amplitude, and frequency content of stress waves introduced into concrete are modified by the presence of interfaces such as cracks, objects, and sections that have impedance.	High level of expertise required to interpret results. Operator requires considerable training to use equipment. Should have knowledge of electronics and should have considerable training in the area of condition survey of concrete structures.	Can operate where only one surface is accessible. Can operate in dry or underwater. Allows one to "see" inside concrete.	Is still in developmental stage. Needs development of measurement criteria. Not presently a standard test method. Digital signal processing can improve interpretation but data must be returned to lab for processing at present.	5, 6, 8, 22, 27, 28
Windsor probe	Estimates of compressive strength, uniformity and quality of concrete.	Probes are gun driven into concrete; depth of penetration converted to estimates of concrete strength by using calibration curves provided by manufacturer.	Simple to operate; can be readily operated in the field with little training.	Equipment is simple, durable and requires little maintenance; useful in assessing the quality and relative strength of concrete; does relatively little damage to specimen.	May not yield accurate estimates of concrete strengths; interpretation of results depends on correlation curves. Difficulty in removing the probes which are often broken and damaging to concrete.	2, 8, 9, 15, 17

References for Table 3.7.2

1. ASTM C597. (1991). "Standard Test Method for Pulse Velocity Through Concrete," 1997 Annual Book of ASTM Standards, Vol. 04.02, American Society for Testing and Materials, West Conshohocken, Pennsylvania. 287–289.
2. ASTM C803. (1996). "Standard Test Method for Penetration of Hardened Concrete," 1997 Annual Book of ASTM Standards, Vol. 04.02, American Society for Testing and Materials, West Conshohocken, Pennsylvania, 389–392.
3. ASTM C805. (1994). "Standard Test Method for Rebound Number of Hardened Concrete," 1997 Annual Book of ASTM Standards, Vol. 04.02, American Society for Testing and Materials, West Conshohocken, Pennsylvania, 393–395.
4. ASTM C1040. (1993). "Standard Test Methods for Density of Unhardened and Hardened Concrete in Place by Nuclear Methods," 1997 Annual Book of ASTM Standards, Vol. 04.02, American Society for Testing and Materials, West Conshohocken, Pennsylvania, 510–513.

5. ASTM D4580. (1992). "Standard Practice for Measuring Delaminations in Concrete Bridge Decks by Sounding," 1997 Annual Book of ASTM Standards, Vol. 04.03, American Society for Testing and Materials, West Conshohocken, Pennsylvania, 471–473.

6. Bay, J. A., and Stokoe, K. H., II. (1990). "Field Determination of Stiffness and Integrity of PCC Slabs Using the SASW Method," *Proc., Non-Destructive Evaluation of Civil Structures and Materials*, edited by B. A. Suprenant, S. Sture, J. L. Noland, and M. P. Schuller, University of Colorado, Boulder, Colorado, October, 71–85.

7. Boundy, C. A. P., and Hondros, G. (1964). "Rapid Field Assessment of Strength of Concrete by Accelerated Curing and Schmidt Rebound Hammer," Journal of The American Concrete Institute, ACI, Vol. 61, Title No. 61-4, January, 77–84.

8. Carino, N. J. (1991). "The Maturity Method," Chapter 5, Handbook on Nondestructive Testing of Concrete, edited by V. M. Malhotra and N. J. Carino, CRC Press, Boca Raton, 101–146.

9. De Brito, J., Branco, F., Batista, A. R., and Cachadinha, M. G. (1989). "Assessment of Existing Structures for their Rehabilitation," *IABSE Symposium, Lisbon, Durability of Structures, IABSE Reports*, Vol. 57/2, September 6–8, 865–870.

10. Galan, A. (1967). "Estimate of Concrete Strength by Ultrasonic Pulse Velocity and Damping Constant," Journal of the American Concrete Institute, ACI, Vol. 64, Title No. 64-59, October, 678–684.

11. Greene, G. W. (1954). "Test Hammer Provides New Method of Evaluating Hardened Concrete," Journal of the American Concrete Institute, ACI, Vol. 51, Title No. 51-11, November, 249–256.

12. Ingvarsson, H. (1987). "Non-Destructive Condition Assessment of Concrete," *IABSE Colloquium, Bergamo, Monitoring of Large Structures and Assessment of their Safety, IABSE Reports*, Vol. 56, September, 65–81.

13. Livingston, R. A. (1990). "Methods for Estimating Chloride Depth Profiles in Reinforced Concrete Using Neutron Probe Data," *Proc., Non-Destructive Evaluation of Civil Structures and Materials*, edited by B. A. Suprenant, S. Sture, J. L. Noland, and M. P. Schuller, University of Colorado, Boulder, Colorado, October, 105–111.

14. Long, A. E., and Murray, A. McC. (1984). "The Pull-off Partially Destructive Test for Concrete," ACI SP-82, In Situ/Nondestructive Testing of Concrete, edited by V. M. Malhotra, American Concrete Institute, Farmington Hills, Michigan, 327–350.

15. Malhotra, V. M. (1976). "Testing Hardened Concrete: Non-destructive Methods," Monograph No. 9, American Concrete Institute, Farmington Hills, Michigan.

16. Malhotra, V. M. (1991). "Surface Hardness Methods," Chapter 1, Handbook on Nondestructive Testing of Concrete, edited by V. M. Malhotra and N. J. Carino, CRC Press, Boca Raton, 1–17.

17. Malhotra, V. M., and Carette, G. G. (1991). "Penetration Resistance Methods," Chapter 2, Handbook on Nondestructive Testing of Concrete, edited by V. M. Malhotra and N. J. Carino, CRC Press, Boca Raton, 19–38.

18. Malhotra, V. M., and Sivasundaram, V. (1991). "Resonant Frequency Methods," Chapter 6, Handbook on Nondestructive Testing of Concrete, edited by V. M. Malhotra and N. J. Carino, CRC Press, Boca Raton, 147–168.

19. Mikhailovsky, L., and Scanlon, A. (1985). "Nondestructive Test Methods for Evaluation of Existing Concrete Bridges," *Proc., 2nd Annual International Bridge Conference*, Pittsburgh, Pennsylvania, Paper No. IBC-85-20, Engineers' Society of Western Pennsylvania, June 17, 18, & 19, 115–120.

20. Mitchell, T. M. (1991). "Radioactive/Nuclear Methods," Chapter 10, Handbook on Nondestructive Testing of Concrete, edited by V. M. Malhotra and N. J. Carino, CRC Press, Boca Raton, 227–252.

21. Montgomery, F. R., Long, A. E., and Basheer, P. A. M. (1989). "Assessing Surface Properties of Concrete by In situ Measurements," *IABSE Symposium, Lisbon, Durability of Structures, IABSE Reports*, Vol. 57/2, September 6–8, 871–876.

22. Muenow, R. (1990). "Concrete Non-destructive Testing-Equipment and Application," *Proc., Non-Destructive Evaluation of Civil Structures and Materials*, edited by B. A. Suprenant, S. Sture, J. L. Noland, and M. P. Schuller, University of Colorado, Boulder, Colorado, October, 341–345.

23. Naik, T. R. (1991). "The Break-Off Test Method," Chapter 4, Handbook on Nondestructive Testing of Concrete, edited by V. M. Malhotra and N. J. Carino, CRC Press, Boca Raton, 83–100.

24. Naik, T. R., and Malhotra, V. M. (1991). "The Ultrasonic Pulse Velocity Method," Chapter 7, Handbook on Nondestructive Testing of Concrete, edited by V. M. Malhotra and N. J. Carino, CRC Press, Boca Raton, 169–188.

25. Sabnis, G. M., and Millstein, L. (1982). "Use of Non-Destructive Methods to Evaluate and Investigate Condition of Buildings and Bridges," *Proc. International Conference on Rehabilitation of Buildings & Bridges Including Investigations*, edited by G. M. Sabnis, Howard University, Washington, D.C., December 21, 22, and 23, 38–40.

26. Samarin, A. (1991). "Combined Methods," Chapter 8, Handbook on Nondestructive Testing of Concrete, edited by V. M. Malhotra and N. J. Carino, CRC Press, Boca Raton, 189–201.

27. Sansalone, M., and Carino, N. J. (1991). "Stress Wave Propagation Methods," Chapter 12, Handbook on Nondestructive Testing of Concrete, edited by V. M. Malhotra and N. J. Carino, CRC Press, Boca Raton, 275–304.

28. Whitehurst, E. A. (1966). "Evaluation of Concrete Properties from Sonic Tests," ACI Monograph No. 2, American Concrete Institute, Farmington Hills, Michigan and The Iowa State University Press, Ames, Iowa.

TABLE 3.7.3. NDT Methods for Determining Physical Conditions of Concrete

Test Method	Purpose of Test	Principle of Operation	User Expertise	Advantage	Limitations	References
Absorption and permeability tests	Determination of surface absorption and permeability.	Attachment of laboratory equipment to concrete surface to create pressure difference and capillary action.	Extensive knowledge required to plan test.	Assessment of concrete quality and curing.	Limited application and experience record.	16
Acoustic emission	Continuous monitoring of structure during service life to detect impending failure; monitoring performance of structure during proof testing.	During crack growth or plastic deformation, the rapid release of strain energy produces acoustic (sound) waves that can be detected by sensors attached to the surface of a test object.	Extensive knowledge required to plan test and to interpret results.	Monitors structural response to applied load; capable of detecting onset of failure; capable of locating source of possible failure; equipment is portable and easy to operate. Good for load tests.	Expensive test to run; can be used only when structure is loaded and when flaws are growing; interpretation of results requires an expert; currently largely confined to laboratory; further work required.	8, 10, 21, 27
Acoustic impact	Used to detect debonds, delaminations, voids, and hairline cracks.	Surface of object is struck with an implement. The frequency through transmission time, and damping characteristics of resulting sound giving an indication of the presence of defects; equipment may vary from simple hammer or drag chain to sophisticated trailer mounted electronic equipment.	Low level of expertise required to use auditory system but electronic system requires training; experience needed for interpreting results.	Portable equipment; easy to perform with auditory system; electronic device requires more equipment.	Geometry and mass of test object influence results; poor discrimination for auditory system; reference standards required for electronic testing.	4, 14, 16
Cover meters/ pachometers	Measure cover, size, and location of reinforcement and metal embedments in concrete or masonry.	Presence of steel in concrete or masonry affects the magnetic field of a probe; closer probe is to steel, the greater the effect.	Moderate; easy to operate; training needed to interpret results.	Portable equipment, good results if concrete is lightly reinforced. Good for location of rebars so as to avoid damage in coring.	Difficult to interpret results if concrete is heavily reinforced or if wire mesh is present. Not reliable for cover over 4 in. (100 mm). Form ties often mistaken for anchors.	8, 9, 27

Electrical potential measurements	Indicating condition of steel rebars in concrete or masonry; indicating the corrosion activity in concrete.	Electrical potential of concrete indicates probability of corrosion.	Moderate, user must be able to recognize problems.	Portable equipment; field measurements readily made; appears to give reliable information.	Information on rate of corrosion not provided; access to rebars required.	8, 12, 16, 25
Electrical resistance measurements	Determination of moisture content of concrete.	Determination of moisture content of concrete is based on the principle that the conductivity of concrete changes with changes in moisture content.	High level of expertise required to interpret results; equipment is easy to use.	Equipment is automated and easy to use.	Equipment very expensive and requires high frequency specialized applications; dielectric properties also depend on salt content and temperature of specimen which poses problems in interpretation of results.	16, 27
Fiber optics	To view portions of a structure which are inaccessible to the eye.	Fiber optic probe consisting of flexible optical fibers; lens and illuminating system is inserted into a crack or drilled hole in concrete; eyepiece is used to view interior to look for flaws such as cracks, voids or aggregate debonds; commonly used to look into areas where cores have been removed or bore holes have been drilled; examination of cavity walls and other holes in masonry.	Equipment is easy to handle and operate.	Gives clear high-resolution images of remote objects. Camera attachment for photos is available. Flexible hose enables multidirectional viewing.	Equipment expensive; many bore holes required to give adequate access. Mortar in masonry walls hinders view.	24
Gamma radiography	Estimating location, size and condition of rebars; voids in concrete; density and thickness.	Based on principle that the rate of absorption of gamma rays is affected by density and thickness of test specimen; gamma rays are emitted from source, penetrate the specimen, exit on opposite side and are recorded on film.	Use of gamma producing isotopes is closely controlled by NRC; equipment must be operated by licensed inspectors.	Internal defects can be detected; applicable to variety of materials; permanent record on film; gamma ray equipment easily portable.	Equipment is very expensive; gamma ray sources are health and safety hazard; require access to both sides of specimen.	8, 9, 16

TABLE 3.7.3. NDT Methods for Determining Physical Conditions of Concrete (*Continued*)

Test Method	Purpose of Test	Principle of Operation	User Expertise	Advantage	Limitations	References
Impact echo	Determination of cracking, cross section properties, delamination and honeycomb.	Monitors surface displacement resulting from interactions of transient stress waves with internal discontinuities.	Moderate level of expertise required.	Simple and efficient technique for interpretation by frequency analysis of displacement waveforms.	The wave speeds must be determined on test object of known thickness.	6, 14
Infrared thermography	Detection of internal flaws, crack growth, delamination, and internal voids.	Flaws detected by using selective infrared frequencies to detect various passive heat patterns which can be identified as belonging to certain defects. Through cracks in concrete and masonry may be detected in cold days.	High level of expertise required to interpret results.	Has potential for becoming a relatively inexpensive and accurate method for detecting concrete defects; can cover large areas quickly.	Requires special skill and equipment.	9, 11, 17, 18, 19, 30, 31
Laser interferometry	Determination of crack initiating and crack propagation and monitoring of the dependent behavior.	A hologram image is created of the object by use of lasers. Fringe patterns represent the displacement field in the object surfaces.	High level of expertise required.	Can monitor time dependent behavior of creep cracks.	High cost. Limited experience with procedure.	15
Nuclear moisture meter	Estimation of moisture content of hardened concrete.	Moisture content in concrete determined based on the principle that materials (such as water) decrease the speed of fast neutrons in accordance with the amount of hydrogen produced in test specimen.	Must be operated by trained and licensed personnel.	Portable; moisture estimates can be made of in-place concrete.	Equipment very sophisticated and expensive; NRC license required to operate; moisture gradients in specimen may give erroneous results.	3, 8
Radar	Detection of substratum voids, delaminations, and embedments. Measurement of thickness of concrete pavements.	Transmitted electromagnetic impulse signals are used for void detection.	High level of expertise required to operate equipment and interpret results.	Expedient method; can locate rebars and voids regardless of depths.	Equipment is expensive; reliability of void detection greatly reduced if reinforcement present; procedure still under development.	5, 7, 9, 11, 13, 17, 18, 20, 21, 29

SEI/ASCE 11-99

Radiographics X-ray and gamma ray	X-Ray-density and internal structure of concrete; location of reinforcement. Gamma ray, location, size and condition of rebars; voids in concrete; density and thickness.	Based on principle that the rate of absorption of X-rays or gamma rays is affected by density and thickness of test specimen; X-rays or gamma rays are emitted from source, penetrate the specimen, exit on opposite side and are recorded on film.	Use of gamma producing isotopes is closely controlled by NRC; gamma equipment must be operated by licensed inspectors.	Internal defects can be detected; applicable to variety of materials; permanent record on film; gamma ray equipment easily portable.	X-ray has limited field application because equipment is heavy and costly; X-ray and gamma ray sources harmful to organic tissue; require access to both sides of specimen.	16, 27
Ultrasonic pulse velocity	Gives indication of strength, uniformity and quality of concrete; internal discontinuities can be located and their size estimated; most widely used stress-wave method for field use.	Operates on principle that stress wave propagation velocity is affected by quality of concrete; pulse waves are introduced in material and the time of arrival measured at the receiving surface with a receiver.	Varying level of expertise required to interpret results. Operator requires a fair degree of training.	Equipment relatively inexpensive and easy to operate; accurate assessment of uniformity and quality. By correlating compressive strength of cores and wave velocity, in situ strength can be estimated.	2, 4, 8, 9, 16, 22, 23, 26, 28, 32
Ultrasonic pulse-echo	Gives indication of uniformity and quality of concrete. Can give location of rebars; voids in concrete; density and thickness.	Operates on principle that original direction, amplitude, and frequency content of stress waves introduced into concrete are modified by the presence of interfaces such as cracks, objects, and sections that have different acoustic impedance.	High level of expertise required to interpret results. Operator requires considerable training to use equipment. Should have knowledge of electronics and should have considerable training in the area of condition survey of concrete structures.	Can operate where only one surface is accessible; can operate in dry or underwater. Allows one to "see" inside concrete.	Is still in developmental stage. Needs development of measurement criteria. Not presently a standard test method. Digital signal processing can improve interpretation but data must be returned to lab for processing at present.	4, 6
Visual examination	a. Evaluation of the surface condition of concrete (finish, roughness, scratches, cracks, color). b. Determining deficiencies in joints. c. Determining differential movements of structure.	Visual examination with or without optical aids, measurement tools, photographic records, or other low cost tools; differential movement determined over long periods of time with surveying methods.	Experience required in order to determine what to look for, what measurements to take, and what follow-up testing to specify.	Generally low cost; rapid evaluation of concrete.	Trained evaluation required; primary evaluation confined to surface of structure.	1, 16

References for Table 3.7.3
1. ACI Committee 201. (1984). "Guide for Making a Condition Survey of Concrete in Service," ACI Designation 201.1R-68 (84), American Concrete Institute, Farmington Hills, Michigan.
2. ACI Committee 228. (1989). "In-Place Methods for Determination of Strength of Concrete," ACI Designation 228.1R-89, American Concrete Institute, Farmington Hills, Michigan.
3. ASTM D3017. (1996). "Standard Test Method for Moisture Content of Soil and Rock in Place by Nuclear Methods (Shallow Depth)," 1997 Annual Book of ASTM Standards, Vol. 04.08, American Society for Testing and Materials, West Conshohocken, Pennsylvania, 295–299.
4. Alexander, A. M., and Hammons, M. I. (1990). "How to Distinguish a Reinforcing Bar From a Void in Concrete Using Non-Destructive Acoustic Techniques," *Proc., Non-Destructive Evaluation of Civil Structures and Materials*, edited by B. A. Suprenant, S. Sture, J. L. Noland, and M. P. Schuller, University of Colorado, Boulder, Colorado, October.
5. Bomar, L. C., Horne, W. F., Brown, D. R., and Smart, J. L. (1988). "Determining Deteriorated Areas in Portland Cement Concrete Pavements Using Radar and Video Imaging," National Cooperative Highway Research Program Report 304, Transportation Research Board, National Research Council, Washington, D.C., December.
6. Carino, N. J. (1984). "Laboratory Study of Flaw Detection in Concrete by the Pulse-Echo Method," In Situ/Nondestructive Testing of Concrete, ACI Publication SP-82, edited by V. M. Malhotra, American Concrete Institute, Farmington Hills, Michigan, 557–579.
7. Clemena, G. G. (1991). "Short-Pulse Radar Methods," Chapter 11, Handbook on Nondestructive Testing of Concrete, edited by V. M. Malhotra and N. J. Carino, CRC Press, Boca Raton, 253–274.
8. Clifton, J. R., and Anderson, E. D. (1981). "Nondestructive Evaluation Methods for Quality Acceptance of Hardened Concrete in Structures," Report NBSIR 80-2163, Center for Building Technology, National Bureau of Standards, U.S. Department of Commerce, Washington, D.C., January.
9. de Vekey, R. C. (1990). "Non-Destructive Evaluation of Structural Concrete: A Review of European Practice and Developments," *Proc., Non-Destructive Evaluation of Civil Structures and Materials*, edited by B. A. Suprenant, S. Sture, J. L. Noland, and M. P. Schuller, University of Colorado, Boulder, Colorado, October, 1–16.
10. Kimura, S., Adachi, I., Hironaka, Y., and Ishibashi, T. (1989). "Acoustic Emission Evaluation of Concrete Structures," *IABSE Symposium, Lisbon, Durability of Structures, IABSE Reports*, Vol. 57/1I, September 6–8, 323–328.
11. Kunz, J. T., and Eales, J. W. (1985). "Evaluation of Bridge Deck Condition by the Use of Thermal Infrared and Ground-Penetrating Radar," *Proc., 2nd Annual International Bridge Conference*, Pittsburgh, Pennsylvania, Paper No. IBC-85-21, Engineers' Society of Western Pennsylvania, June 17, 18, & 19, 121–127.
12. Lauer, K. R. (1991). "Magnetic/Electrical Methods," Chapter 9, Handbook on Nondestructive Testing of Concrete, edited by V. M. Malhotra and N. J. Carino, CRC Press, Boca Raton, 203–225.
13. Lim, M. K., and Olson, C. A. (1990). "Use of Non-Destructive Impulse Radar in Evaluating Civil Engineering Structures," *Proc., Non-Destructive Evaluation of Civil Structures and Materials*, edited by B. A. Suprenant, S. Sture, J. L. Noland, and M. P. Schuller, University of Colorado, Boulder, Colorado, October, 167–176.
14. Limaye, H. S., and Klein, G. J. (1990). "Investigation of Concrete Arch Bridges Using the Impact-Echo Method," *Proc., Non-Destructive Evaluation of Civil Structures and Materials*, edited by B. A. Suprenant, S. Sture, J. L. Noland, and M. P. Schuller, University of Colorado, Boulder, Colorado, October, 221–231.
15. Maji, A. K. (1990). "Civil Engineering Applications of Laser Interferometry," *Proc., Non-Destructive Evaluation of Civil Structures and Materials*, edited by B. A. Suprenant, S. Sture, J. L. Noland, and M. P. Schuller, University of Colorado, Boulder, Colorado, October, 155–166.
16. Manning, D. G. (1985). "Detecting Defects and Deterioration in Highway Structures," National Cooperative Highway Research Program, Synthesis of Highway Practice 118, Transportation Research Board, National Research Council, Washington, D.C., July.

17. Manning, D. G., and Masliwec, T. (1990a). "Application of Radar and Thermography to Bridge Deck Condition Surveys," Bridge Management—Inspection, Maintenance, Assessment and Repair, edited by J. E. Harding, G. A. R. Parke, and M. J. Ryall, Elsevier Applied Science, London and New York, 305–317.

18. Manning, D. G., and Masliwec, T. (1990b). "Operational Experience Using Radar and Thermography for Bridge Deck Condition Survey," *Proc., Non-Destructive Evaluation of Civil Structures and Materials*, edited by B. A. Suprenant, S. Sture, J. L. Noland, and M. P. Schuller, University of Colorado, Boulder, Colorado, October, 233–244.

19. Maser, K. R., and Roddis, W. M. K. (1988). "Infrared Thermography and Radar for Bridge Deck Assessment," *Proc., 5th Annual International Bridge Conference*, Pittsburgh, Pennsylvania, Paper No. IBC-88-24, Engineers' Society of Western Pennsylvania, June 13, 14, & 15, 120–125.

20. Mast, J. E., Lee, H., Chew, W. C., and Murtha, J. P. (1990). "Pulse-Echo Holographic Techniques for Microwave Subsurface NDE," *Proc., Non-Destructive Evaluation of Civil Structures and Materials*, edited by B. A. Suprenant, S. Sture, J. L. Noland, and M. P. Schuller, University of Colorado, Boulder, Colorado, October, 177–191.

21. Mindess, S. (1991). "Acoustic Emission Methods," Chapter 14, Handbook on Nondestructive Testing of Concrete, edited by V. M. Malhotra and N. J. Carino, CRC Press, Boca Raton, 317–333.

22. Nazarian, S. (1990). "Detection of Deterioration Within and Beneath Concrete Pavements with Sonic and Ultrasonic Surface Waves," *Proc., Non-Destructive Evaluation of Civil Structures and Materials*, edited by B. A. Suprenant, S. Sture, J. L. Noland, and M. P. Schuller, University of Colorado, Boulder, Colorado, October, 391–406.

23. Olson, L. D. (1990). "NDE of Structural Concrete with Stress Waves," *Proc., Non-Destructive Evaluation of Civil Structures and Materials*, edited by B. A. Suprenant, S. Sture, J. L. Noland, and M. P. Schuller, University of Colorado, Boulder, Colorado, October, 61–70.

24. Olympus Industrial Fiberoptics Division. (1986). "Focus on Inspection," Vol. IV, No. 3, December.

25. Otto, G., Chew, W. C., and Young, J. F. (1990). "A Large Open-ended Coaxial Probe for Dielectric Measurements of Cements and Concretes," *Proc., Non-Destructive Evaluation of Civil Structures and Materials*, edited by B. A. Suprenant, S. Sture, J. L. Noland, and M. P. Schuller, University of Colorado, Boulder, Colorado, October, 193–209.

26. Popovics, J. S. (1990). "Are Advanced Ultrasonic Techniques Suitable for Concrete?—An Exploratory Investigation." *Proc., Non-Destructive Evaluation of Civil Structures and Materials*, edited by B. A. Suprenant, S. Sture, J. L. Noland, and M. P. Schuller, University of Colorado, Boulder, Colorado, October, 327–339.

27. Roddis, W. M. K. (1987). "Concrete Bridge Deck Condition Assessment: Traditional and Innovative Inspection Technologies," *Proc., Infrastructure—Repairs and Inspection*. Sponsored by the Structural Division of the American Society of Civil Engineers in Conjunction with the ASCE Convention in Atlantic City, New Jersey, edited by M. J. Shah, ASCE, Reston, Virginia, April 29, 73–85.

28. Sabnis, G. M., Kelishami, R., and Millstein, L. (1990). "Delamination Detection in Concrete Bridge Decks Using Non-Destructive Test Method," *Proc., Non-Destructive Evaluation of Civil Structures and Materials*, edited by B. A. Suprenant, S. Sture, J. L. Noland, and M. P. Schuller, University of Colorado, Boulder, Colorado, October, 371–389.

29. Sarnes, F. W., Jr., Murphy, V. J., and Sankey, F. C. (1985). "Integration of Geophysical Techniques and Direct Inspection for Bridge Deck Evaluation: Market Street Bridge," *Proc., 2nd Annual International Bridge Conference*, Pittsburgh, Pennsylvania, Paper No. IBC-85-22, Engineers' Society of Western Pennsylvania, June 17, 18, & 19, 128–134.

30. Schaufler, E. R. (1984). "Thermographic Diagnosis of Bridge Deck Delamination," *Proc., 1st International Bridge Conference*, Pittsburgh, Pennsylvania, Paper No. IBC-84-38, Engineers' Society of Western Pennsylvania, June 4, 5, & 6, 247–251.

31. Weil, G. J. (1991). "Infrared Thermographic Techniques," Chapter 13, Handbook on Nondestructive Testing of Concrete, edited by V. M. Malhotra and N. J. Carino, CRC Press, Boca Raton, 305–316.

32. Wiberg, U. (1989). "Ultrasonic Testing of Concrete," *IABSE Symposium, Lisbon, Durability of Structures, IABSE Reports*, Vol. 57/1, September 6–8, 317–322.

TABLE 3.7.4. Test Methods for Determining Chemical and Physical Properties of Reinforcing Steel, Pre-Tensioning Steel, and Post-Tensioning Steel on Metal Specimens

Test Method	Purpose of Test	Principle of Operation	User Expertise	Advantages	Limitations	References
Bend test	Determination of ductility of metal specimen and resistance to cracking.	Bend specimen at room temperature through 180° without cracking on outside of bent portion.	Low. Can be performed by any qualified laboratory.	Gives fast accurate information. Low cost.	Samples must be removed from structure for testing.	2, 5, 6, 9, 10, 11, 12, 13, 14, 15
Breaking strength	Determination of breaking strength of finished strand after stress relieving and thermal treatment. Includes determination of elongation, load at 1% extension, and yield strength.	Testing and breaking of strands in an extensometer and production of load elongation curve.	Standard laboratory test. Moderate expertise required.	Provides accurate and useful information.	Sample must be removed from structure. These are usually destructive.	6, 7, 20
Chemical analysis	To obtain compositions of metal in order to identify type of metal and to establish such characteristics as weldability, corrosion resistance, ductility and other mechanical* characteristics that can be established by the knowledge of chemical compositions.	A sample of metal taken from the structure and is chemically analyzed to establish composition of metal.	Moderate; can be performed by any qualified chemical analysis laboratory.	Are the common tests to provide data on all chemical compositions of metal. Data can be used to evaluate corrosion resistance, weldability and some mechanical properties. Can be used to establish identity of metal. Although mainly a laboratory test, field testing can also be performed for spot testing.	Can be destructive if samples are removed indiscriminately from structure. Experience needed to conduct the tests. Care should be taken in sampling to ensure that specimen is not contaminated by other chemicals, water, and oil.	8, 11, 12, 15, 16, 17, 20, 21
Deformation requirements	Determination of minimum average height and maximum and minimum spacing of deformations.	Standard measurements.	Low; can be performed outside laboratory.	Provides useful and required information.	Usually accomplished by in situ tests but samples may be taken from structure. These may be destructive.	9, 10, 11, 12, 13, 14, 15

SEI/ASCE 11-99

Film thickness	Determination of film thickness of epoxy-coated reinforcing steel bars.	Use of a nondestructive-type thickness gauge.	Low; can be performed outside laboratory.	Provides useful and required information.	Samples must be taken from structure. These may be destructive.	1, 19, 24
Strength of connections (in welded or clipped deformed bar mats)	Determination of specified static load without loosening.	Tests of connections against slipping are performed on an assembled mat by means of a spring balance.	Standard laboratory test.	Provides useful and required information.	Samples must be taken from structure. These may be destructive.	4
Stress relaxation test	Determination of time-dependent decrease in stress in a specimen subjected to uniaxial constant tension strain.	Tension of specimen in a testing machine under constant temperature condition and recording strain incrementally, and after maintaining strain for specified time, measuring changes.	Standard laboratory test. Moderately high expertise required.	Provides useful and required information.	Samples must be taken from structure. These may be destructive.	7, 8, 20, 22
Tensile test	To obtain yield strength, yield point, tensile strength, elongation and reduction of area. Sometimes a stress-strain or load-elongation curve is also recorded during the test that can be useful in evaluating ductility and modulus of elasticity.	A standard metal specimen is subjected to axial tension and major mechanical properties are measured.	Low; can be performed by any qualified laboratory.	Gives fast and accurate results on mechanical properties of metals.	Samples must be removed from the structure for testing. These could be destructive.	5, 6, 8, 9, 10, 11, 12, 13, 14, 15, 16
Weight of coating (galvanized products)	Determination of coating weight by either stripping specimens or use of a magnetic thickness gauge.	Determination of weight of stripped specimens or measuring required coating thickness.	Standard laboratory test. Moderate expertise required.	Provides required information.	Usually an in-situ test. When samples are removed from structure, these could be destructive.	3, 18, 23
Weld shear test	Determination of minimum average shear values in both transverse and longitudinal directions of welded wire fabric.	Testing of fabric in a specially prepared jig utilizing a standard tension testing machine.	Standard laboratory test. Moderate expertise required.	Provides required information.	Samples must be removed from structure. These could be destructive.	5

TABLE 3.7.4. Test Methods for Determining Chemical and Physical Properties of Reinforcing Steel, Pre-Tensioning Steel, and Post-Tensioning Steel on Metal Specimens (*Continued*)

Test Method	Purpose of Test	Principle of Operation	User Expertise	Advantages	Limitations	References
Wire diameter	Determination of nominal diameter and conformance with permissible variations of plain steel wire for concrete reinforcement.	Measurement of wire diameters.	Low: may be performed outside laboratory.	Provides required information.	Usually an in situ test. When samples are taken from structure, these may be destructive.	2, 5

*Chemical analysis is not usually a reliable indicator of mechanical properties.

References for Table 3.7.4

1. U.S. Federal Highway Administration, FHWA-RD-74-18. (1974). "Nonmetallic Coatings for Concrete Reinforcing Bars," Federal Highway Administration Report FHWA-RD-74-18. U.S. Department of Transportation, Washington, D.C., February.
2. ASTM A82. (1995a). "Standard Specification for Steel Wire, Plain, for Concrete Reinforcement," 1997 Annual Book of ASTM Standards, Vol. 01.04, American Society for Testing and Materials, West Conshohocken, Pennsylvania, 119–122.
3. ASTM A90. (1995a). "Test Method for Weight (Mass) of Coating on Iron and Steel Articles with Zinc or Zinc-Alloy Coatings," 1997 Annual Book of ASTM Standards, Vol. 01.06, American Society for Testing and Materials, West Conshohocken, Pennsylvania, 1–4.
4. ASTM A184/A184M. (1996). "Standard Specification for Fabricated Deformed Steel Bar Mats for Concrete Reinforcement," 1997 Annual Book of ASTM Standards, Vol. 01.04, American Society for Testing and Materials, West Conshohocken, Pennsylvania, 132–134.
5. ASTM A185. (1994). "Standard Specification for Steel Welded Wire Fabric, Plain, for Concrete Reinforcement," 1997 Annual Book of ASTM Standards, Vol. 01.04, American Society for Testing and Materials, West Conshohocken, Pennsylvania, 135–139.
6. ASTM A370. (1996). "Standard Test Methods and Definitions for Mechanical Testing of Steel Products," 1997 Annual Book of ASTM Standards, Vol. 01.04, American Society for Testing and Materials, West Conshohocken, Pennsylvania, 164–209.
7. ASTM A416. (1996). "Standard Specification for Steel Strand, Uncoated Seven-Wire for Prestressed Concrete," 1997 Annual Book of ASTM Standards, Vol. 01.04, American Society for Testing and Materials, West Conshohocken, Pennsylvania, 214–217.
8. ASTM A421. (1991). "Standard Specification for Uncoated Stress-Relieved Steel Wire for Prestressed Concrete," 1997 Annual Book of ASTM Standards, Vol. 01.04, American Society for Testing and Materials, West Conshohocken, Pennsylvania, 218–220.
9. ASTM A496. (1995a). "Standard Specification for Steel Wire, Deformed, for Concrete Reinforcement," 1997 Annual Book of ASTM Standards, Vol. 01.04, American Society for Testing and Materials, West Conshohocken, Pennsylvania, 225–229.

10. ASTM A497. (1995). "Standard Specification for Steel Welded Wire Fabric, Deformed, for Concrete Reinforcement," 1997 Annual Volume of ASTM Standards, Vol. 01.04, American Society for Testing and Materials, West Conshohocken, Pennsylvania, 230–234.
11. ASTM A615. (1996a). "Standard Specification for Deformed and Plain Billet-Steel Bars for Concrete Reinforcement," 1997 Annual Book of ASTM Standards, Vol. 01.04, American Society for Testing and Materials, West Conshohocken, Pennsylvania, 301–305.
12. ASTM A615M. (1996a). "Standard Specification for Deformed and Plain Billet-Steel Bars for Concrete Reinforcement (Metric)," 1997 Annual Book of ASTM Standards, Vol. 01.04, American Society for Testing and Materials, West Conshohocken, Pennsylvania, 301–305.
13. ASTM A616/A616M. (1996a). "Standard Specification for Rail-Steel Deformed and Plain Bars for Concrete Reinforcement," 1997 Annual Volume of ASTM Standards, Vol. 01.04, American Society for Testing and Materials, West Conshohocken, Pennsylvania, 306–310.
14. ASTM A617/A617M. (1996a). "Standard Specification for Axle-Steel Deformed and Plain Bars for Concrete Reinforcement," 1997 Annual Book of ASTM Standards, Vol. 01.04, American Society for Testing and Materials, West Conshohocken, Pennsylvania, 310–313.
15. ASTM A706/A706M. (1996b). "Standard Specification for Low-Alloy Steel Deformed and Plain Bars for Concrete Reinforcement," 1997 Annual Book of ASTM Standards, Vol. 01.04, American Society for Testing and Materials, West Conshohocken, Pennsylvania, 340–344.
16. ASTM A722. (1995). "Standard Specification for Uncoated High-Strength Steel Bar for Prestressing Concrete," 1997 Annual Book of ASTM Standards, Vol. 01.04, American Society for Testing and Materials, West Conshohocken, Pennsylvania, 355–358.
17. ASTM A751. (1995). "Standard Test Methods, Practices, and Terminology for Chemical Analysis of Steel Products," 1997 Annual Book of ASTM Standards, Vol. 01.03, American Society for Testing and Materials, West Conshohocken, Pennsylvania, 385–389.
18. ASTM A767/A767M. (1995). "Standard Specification for Zinc-Coated (Galvanized) Steel Bars for Concrete Reinforcement," 1997 Annual Book of ASTM Standards, Vol. 01.04, American Society for Testing and Materials, West Conshohocken, Pennsylvania, 393–396.
19. ASTM A775/A775M. (1996). "Standard Specification for Epoxy-Coated Reinforcing Steel Bars," 1997 Annual Book of ASTM Standards, American Society for Testing and Materials, West Conshohocken, Pennsylvania, 406–412.
20. ASTM A779. (1995). "Standard Specification for Steel Strand, Seven-Wire, Uncoated, Compacted, Stress-Relieved for Prestressed Concrete," 1997 Annual Book of ASTM Standards, Vol. 01.04, American Society for Testing and Materials, West Conshohocken, Pennsylvania, 413–415.
21. ASTM E30. (1989). "Standard Test Methods for Chemical Analysis of Steel, Cast Iron, Open-Hearth Iron, and Wrought Iron," 1994 Annual Book of ASTM Standards, Vol. 03.05, American Society for Testing and Materials, West Conshohocken, Pennsylvania, 39–50.
22. ASTM E328. (1996). "Standard Methods for Stress Relaxation Tests for Materials and Structures," 1997 Annual Book of ASTM Standards, Vol. 03.01, American Society for Testing and Materials, West Conshohocken, Pennsylvania, 349–360.
23. ASTM E376. (1996). "Standard Practice for Measuring Coating Thickness by Magnetic-Field or Eddy-Current (Electromagnetic) Test Methods," 1997 Annual Book of ASTM Standards, Vol. 03.03, American Society for Testing and Materials, West Conshohocken, Pennsylvania, 132–135.
24. ASTM G12. (1992). "Standard Test Method for Nondestructive Measurement of Film Thickness of Pipeline Coatings on Steel," 1997 Annual Book of ASTM Standards, Vol. 06.02, American Society for Testing and Materials, West Conshohocken, Pennsylvania, 652–653.

TABLE 3.7.5. NDT Methods for Determining Physical Conditions of Reinforcing Steel, Pre-Tensioning Steel, and Post-Tensioning Steel

Test Method	Purpose of Test	Principle of Operation	User Expertise	Advantages	Limitations	References
Acoustic emission	Determination of degree of debonding of reinforcing steel.	During fracturing of concrete around reinforcing steel bar deformations, the release of strain energy produces acoustic waves that can be detected by sensors attached to the surface of the test object.	Extensive knowledge and expertise required.	Equipment is portable and easy to operate. Can be used for long term monitoring. Method is nondestructive.	Expensive. Requires complex equipment. Quantitative results cannot be obtained.	17
Copper wire sensor	Determination of change in strain (stress) state of prestressing steel.	The change in capacity between different prestressing tendons is recorded under changes in load. The change in strain is mutually proportional.	Extensive knowledge and expertise required.	Easy to operate. Can be used for long term monitoring.	Installation of copper wires expensive and can usually only be accomplished at tendons close to surface.	23
Coring stress relief	Determination of average stress field around prestressed reinforcement.	Based on the stress-relief which occurs during removal of concrete cores and measuring resulting strains.	Extensive knowledge and expertise required.	Measures stresses fairly accurately and method can be used with a good degree of confidence. Quick and simple.	Expensive. Local anomalies can affect results.	5
Electrochemical impedance	Monitoring corrosion of reinforcing steel.	Measurement of the impedance due to the electrochemistry of the active corrosion cell.	Moderate knowledge and expertise required.	Can determine instantaneous corrosion rate in a quick and easy manner.	Approximate technique. Many measurements required to determine overall pattern of corrosion.	3, 4, 13, 14
Flat jack	Direct measurement of stresses in concrete adjacent to reinforcing steel.	A slot is cut into concrete and a flat jack is installed into the slot and initial displacement restored. Strain (stress) is measured to restore displacement field.	Moderate knowledge and expertise required.	Accurate. Fairly simple concept to apply and use.	Usually limited to measuring uni-axial stresses although other states of stress can be measured by use of several slots to construct a Mohr Circle.	2, 19
Galvanic current	Study effect of chloride concentration.	Galvanic current measured of electrochemical corrosion cell.	Moderate knowledge and expertise required.	Accurate measurement of corrosion rates.	Many measurements required to establish pattern of corrosion.	12

Half-cell potentials	Evaluation of potential measurements in reinforcing steel which are a measure of active corrosion.	A copper-copper sulfate half cell is attached—one lead to reinforcing steel and other to concrete surfaces which is moved around to record readings of potential in rebars. Other types of half cells are also available. The rate of change of potential gradient is the indicator of corrosion.	Low level of expertise required.	Inexpensive; easy to use; fairly fast.	Prewetting of concrete surface required. Connection to reinforcing steel involves minor demolition. There is considerable natural variation in the results. The method does not indicate the corrosion rate.	1, 6, 7, 8, 13, 16
Magnetic field	Determination of fractures in reinforcing strand and deterioration of reinforcing steel.	Application of a steady-state magnetic field and use of a scanning magnetic field sensor to detect perturbances in the applied field caused by anomalies.	Moderate knowledge and expertise required.	Good overall sensitivity to loss of section and sensitively to fractures.	Expensive equipment required. Method still experimental.	13
Magnetic permeability (Tensiomag)	Stresses in reinforcing steel and prestressing steel can be measured and monitored with time.	The incremental magnetic permeability is a measure of the gradient of a minor hysterisis loop. Steel tendons are magnetized and state of magnetization measured.	Expert knowledge and experience required.	Accurate measurement of stress in reinforcement and prestressing steel.	Measurements limited to parallel reinforcement or strand configuration and also limited by size of strand.	18
Magnoelastic force (Pressductor)	Stresses in prestressing steel can be determined.	Magnetic properties of strands changes under influence of forces. The length of wire is changed by magnetizing it and change in length is measured.	Expert knowledge and experience required.	Can be installed at any position along tendon.	Force measurement is dependent upon the chemical composition and grain structure of tendon.	10

TABLE 3.7.5. NDT Methods for Determining Physical Conditions of Reinforcing Steel, Pre-Tensioning Steel, and Post-Tensioning Steel *(Continued)*

Test Method	Purpose of Test	Principle of Operation	User Expertise	Advantages	Limitations	References
Optical fiber sensors	Determination of change in strain (stress) of prestressing steel.	A specially coated optical fiber is covered with a thin wire in a spiral pattern. The path of the spiral changes with axial stretching and this presses on the optical fiber generating microbendings. These influence permeability of light which is proportional to strain.	Extensive knowledge and expertise required.	Easy to operate. Can be used for long term monitoring.	Installation of sensor into concrete is locally semi-destructive and the methods limited to accessible tendons.	23
Polarization resistance	Measurement of corrosion rate in reinforcing steel.	The corrosion current due to electrochemical corrosion cell is measured.	Extensive knowledge and expertise required.	Accurate determination of corrosion rate of reinforcing steel.	Portable computer system required for control. Expensive.	4, 9, 13, 15
Radioscopy	Assessment of quality of grouting in prestressed tendons.	Use of miniaturized linear accelerator.	Extensive knowledge and expertise required.	Useful method providing required information.	Expensive. Considerable equipment required.	11
Suspension cable and stayed cable wire fractures by acoustic survey	Determination of fractured wires in cables.	Acoustic sensors installed on to cables. Waves from accelerometer monitored remotely.	Extensive knowledge and expertise required.	Method has been field tested on a number of bridges and provides reliable results when compared to destructive tests.	Information limited by number of sensors installed. Expensive. Only future breaks can be determined.	20
Suspension cable and stayed cable wire fractures by magnetic perturbation	Determination of fractured wires in cable.	Cables are scanned magnetically by a mobile unit consisting of an electromagnet and sensor. Flaws in wire produce magnetic perturbations from which flaw size depth and location can be determined.	Extensive knowledge and expertise required.	Method has been field tested in a number of bridges as well as in laboratory. Can detect fractures reliably.	Limited equipment available. Expensive. Still experimental.	22
X-ray permeability	Measurement of corrosion rate of reinforcing steel.	The diffusion of ferrous ions from surface of steel undergoing corrosion is identified by an electron probe.	Extensive knowledge and expertise required.	Method is quantitative.	Still in experimental stage.	21

References for Table 3.7.5

1. ASTM C876. (1991). "Standard Test Method for Half-Cell Potentials of Uncoated Reinforcing Steel in Concrete," 1997 Annual Book of ASTM Standards, Vol. 04.02, American Society for Testing and Materials, West Conshohocken, Pennsylvania. 426–431.
2. Abdunur, C. (1982). "Direct Measurement of Stresses in Concrete Structures," IABSE Symposium, Washington, D.C., Maintenance, Repair and Rehabilitation of Bridges, Final Report, IABSE Reports, Vol. 39, pp. 39–44.
3. Aguilar, A., Sagues, A. A., and Powers, R. G. (1990). "Corrosion Measurements of Reinforcing Steel in Partially Submerged Concrete Slabs," Symposium on Corrosion Rates of Steel in Concrete, ASTM STP 1065, edited by N. S. Berke, V. Chaker, and D. Whiting, American Society for Testing and Materials, West Conshohocken, Pennsylvania. 66–85.
4. Andrade, C., Castelo, V., Alonso, C., and Gonzlez, J. A. (1986). "The Determination of the Corrosion Rate of Steel Embedded in Concrete by the Polarization Resistance and AC Impedance Methods," Corrosion Effect of Stray Currents and the Techniques for Evaluating Corrosion of Rebars in Concrete, ASTM STP 906, edited by V. Chaker, American Society for Testing and Materials, West Conshohocken, Pennsylvania. 43–63.
5. Brookes, C. L., Buchner, S. H., and Mehrkar-Asl. S. (1990). "Assessment of Stresses in Post-Tensioned Concrete Bridges," Bridge Management—Inspection, Maintenance, Assessment and Repair, edited by J. E. Harding, G. A. R. Parke, and M. J. Ryall, Elsevier Applied Science, London and New York. 439–446.
6. Broomfield. J. P., Langford, P. E., and Ewins, A. J. (1990). "The Use of a Potential Wheel to Survey Reinforced Concrete Structures," Symposium on Corrosion Rates of Steel in Concrete, ASTM STP 1065, edited by N. S. Berke, V. Chaker, and D. Whiting, American Society for Testing and Materials, West Conshohocken, Pennsylvania. 157–173.
7. Cook, A. R., and Radtke, S. F. (1977). "Recent Research on Galvanized Steel for Reinforcement of Concrete," Chloride Corrosion of Steel in Concrete, ASTM STP 629, edited by D. E. Tonini and S. W. Dean, American Society for Testing and Materials, West Conshohocken, Pennsylvania. 51–60.
8. Elsener, B., and Bohni, H. (1990). "Potential Mapping and Corrosion of Steel in Concrete," Symposium on Corrosion Rates of Steel in Concrete, ASTM STP 1065, edited by N. S. Berke, V. Chaker, and D. Whiting, American Society for Testing and Materials, West Conshohocken, Pennsylvania. 143–156.
9. Escalante, E. and Ito, S. (1990). "Measuring the Rate of Corrosion of Steel in Concrete," Symposium on Corrosion Rates of Steel in Concrete, ASTM STP 1065, edited by N. S. Berke, V. Chaker, and D. Whitney, American Society for Testing and Materials, West Conshohocken, Pennsylvania. 86–102.
10. Gimmel, B. (1989). "Magnetoelastic Force Measurement in Prestressed Concrete," IABSE Symposium, Lisbon, Durability of Structures, IABSE Reports, Vol. 57/1, September 6–8, 329–334 (in German).
11. Guinez, R., Chatelain, J., and Chevrier, J.-P. (1989). "Radioscopy of Prestressed Concrete Bridges," IABSE Symposium, Lisbon, Durability of Structures, IABSE Reports, Vol. 57/1, September 6–8, 347–352 (in French).
12. Jang, J. W., and Iwasaki, I. (1991). "Rebar Corrosion under Simulated Concrete Conditions using Galvanic Current Measurements," Proc., 70th Annual Meeting, Paper No. 910155, Transportation Research Board, Washington, D.C. January.
13. Lauer, K. R. (1991). "Magnetic/Electrical Methods," Chapter 9, Handbook on Nondestructive Testing of Concrete, edited by V. M. Malhotra and N. J. Carino, CRC Press, Boca Raton, 203–225.
14. Lemoine, L., Wenger, F., and Galland, J. (1990). "Study of the Corrosion of Concrete Reinforcement by Electrochemical Impedance Measurement," Symposium on Corrosion Rates of Steel in Concrete, ASTM STP 1065, edited by N. S. Berke, V. Chaker, and D. Whiting, American Society for Testing and Materials, West Conshohocken, Pennsylvania. 118–133.
15. Locke, C. E., and Siman, A. (1980). "Electrochemistry of Reinforcing Steel in Salt-Contaminated Concrete," Symposium on Corrosion of Reinforcing Steel in Concrete, ASTM STP 713, edited by D. E. Tonini and J. M. Gaidis, American Society for Testing and Materials, West Conshohocken, Pennsylvania. 3–16.
16. Matsuoka, K., Kihira, H., Ito, S., and Murata, T. (1990). "Corrosion Monitoring for Reinforcing Bars in Concrete," Symposium on Corrosion Rates of Steel in Concrete, ASTM STP 1065, edited by N. S. Berke, V. Chaker, and D. Whiting, American Society for Testing and Materials, West Conshohocken, Pennsylvania. 103–117.
17. Mindess, S. (1991). "Acoustic Emission Methods," Chapter 14, Handbook on Nondestructive Testing of Concrete, edited by V. M. Malhotra and N. J. Carino, CRC Press, Boca Raton, 317–333.
18. Orsat, P., and Bertel, J.-C. (1988). "Measurement of Forces Actually Applied on Rebars Anchor-Bolts or Strands with Tensiomag," Proc., 5th Annual International Bridge Conference, Pittsburgh, Pennsylvania, Paper No. IBC-88-14, Engineers' Society of Western Pennsylvania, June 13, 14, & 15, 51–56.
19. Overman, T. R., and Hanson, N. W. (1986). "Assessment of Prestressed Concrete Stresses by Direct Stress Measurement," Proc., 3rd Annual International Bridge Conference, Pittsburgh, Pennsylvania, Paper No. IBC-86-18, Engineers' Society of Western Pennsylvania, June 2, 3, & 4, 118–122.
20. Robert, J. L., and Brachet-Rolland, M. (1982). "Survey of Structures by Using Acoustic Emission Monitoring," IABSE Symposium, Washington, D.C., Maintenance, Repair and Rehabilitation of Bridges, Final Report, IABSE Reports, Vol. 39, 33–38.
21. Skoulikidis, T., Marinakis, D., and Batis, G. (1985). "X-Ray Permeability of Corrosion Products as a Measure of the Rate of Corrosion of Rebars and of Prediction of Concrete Cracking," Corrosion Effect of Stray Currents and the Techniques for Evaluating Corrosion of Rebars in Concrete, ASTM STP 906, edited by V. Chaker, American Society for Testing and Materials, West Conshohocken, Pennsylvania. 108–117.
22. Teller, C. M., Matzkanin, G. A., and Shuler, S. A. (1990). "NDE of Suspension Bridge Cables Using a Recently Developed Magnetic Perturbation Cable (MPC) Inspection System," Proc., Non-Destructive Evaluation of Civil Structures and Materials, edited by B. A. Suprenant, S. Sture, J. L. Noland, and M. P. Schuller, University of Colorado, Boulder, Colorado, October, 277–295.
23. Wolff, R., and Miesseler, H.-J. (1990). "Prestressing with Fibre Composite Materials and Monitoring of Bridges with Sensors," Bridge Management—Inspection, Maintenance, Assessment and Repair, edited by J. E. Harding, G. A. R. Parke, and M. J. Ryall, Elsevier Applied Science, London and New York. 395–402.

TABLE 3.7.6. Test Methods for Determining Chemical and Physical Properties of Steel on Metal Specimens

Test Method	Purpose of Test	Principal of Operation	User Expertise	Advantages	Limitations	References
Chemical analysis tests of metallic materials and connectors	To obtain compositions of metal in order to identify type of metal and to establish such characteristics as weldability, corrosion resistance, ductility and other mechanical characteristics that can be established by the knowledge of chemical compositions.	A sample of metal or a sample connector, taken from the structure, is chemically analyzed to establish composition of metal.	Moderate, can be performed by any qualified chemical analysis laboratory.	Common tests to provide data on most chemical compositions of metal. Data may in some instances be used to evaluate corrosion resistance, weldability and some mechanical properties. Can be used to establish identity of metal. Although mainly a laboratory test, field testing can also be performed for spot testing.	Can be destructive if samples are removed indiscriminately from structure. Experience needed to conduct the tests. Care should be taken in sampling to ensure that specimen is not contaminated by other chemicals, water, and oil.	3, 10
Compression test	To obtain modulus of elasticity, yield strength, yield point and compressive strength.	A test specimen is taken from the structure and prepared. It is subjected to increasing axial compressive load with both load and strain monitored.	Moderate; can be performed by any qualified testing laboratory.	Is used to determine mechanical properties; tension test is more common.	Buckling of specimen must be controlled and barreling of the end section may result in nonuniform transverse deformation.	6
Fastener mechanical properties tests	To determine mechanical properties of threaded fasteners, washers and rivets including product hardness, proof load and axial tension.	Samples of fasteners are required for testing.	Moderate expertise required to perform tests.	High degree of accuracy since fasteners are usually available. Tests can be considered nondestructive.	Expensive. Information limited by number of samples tested.	21

SEI/ASCE 11-99

Test	Purpose	Description	Advantages	Disadvantages	Refs.
Fatigue properties tests	To obtain fatigue characteristics of metals that can be used in establishing total or remaining life of metal subjected to a given history of cyclic stress or strain. Common fatigue properties in high cycle fatigue are S-N curve, e-N curve, endurance limit, and constant life diagram. The major properties in low-cycle fatigue are hysteresis loops, cyclic strain hardening exponent and the cyclic strength coefficient.	Several tests are conducted to establish fatigue properties of metals. In these tests standard specimens are subjected to axial, torsion, or bending cyclic strain or stress. In high cycle low amplitude fatigue, the amplitude of stress or strain is lower than yield point and a large number of cycles are required to cause fatigue damage. In low-cycle high amplitude fatigue, the amplitude of strain is higher than yield strain. In this case a small number of cycles can cause fracture due to low-cycle fatigue.	Provide useful information on fatigue resistance of metals. Experience is required to conduct the tests and interpret the test results.	High costs; difficulty in obtaining nondestructive samples from structure; results sometimes cannot be directly applied to the field conditions unless service conditions are the same as test conditions.	14, 16, 17, 20
Fracture tests	To obtain fracture properties such as plane strain fracture toughness, R-curve, J-integral, nil-ductility transition temperature, and qualitative information on fracture behavior of metals.	Various tests are conducted to establish various fracture properties of metals. In these tests standard notched specimens are subjected to monotonic or impact loading causing fracture of specimen through the notch. Using load-deflection, load-crack tip opening displacement and other recordings during the test, fracture properties of metal are established. The properties can be used to estimate resistance of metal to fracture.	Moderate cost; gives quantitative measures of fracture toughness to be used in analysis. Qualified laboratories can perform the test. Experience is required to conduct the test.	Difficulties in obtaining nondestructive samples from structure. In some cases there is no advance assurance that a valid test result will be obtained. Size limitations on specimen exist that in some instances the standard specimen cannot be fabricated. Careful specimen preparation is needed. Test results are sensitive to operation and are affected by residual stresses.	12, 13, 15, 18, 19

GUIDELINE FOR STRUCTURAL CONDITION ASSESSMENT OF EXISTING BUILDINGS

TABLE 3.7.6. Test Methods for Determining Chemical and Physical Properties of Steel on Metal Specimens (*Continued*)

Test Method	Purpose of Test	Principal of Operation	User Expertise	Advantages	Limitations	References
Hardness tests	A hard steel ball or diamond shape penetrator is forced into the material using standard procedures. By measuring the geometry of the indentation mark left on the material, hardness number is established. Several hardness numbers are defined and used in industry. Brinell and Rockwell hardness numbers are used frequently to define hardness of metals.	To evaluate resistance to permanent or plastic deformations. To obtain quick but approximate value of tensile strength. Also used to evaluate quality level and uniformity of heat treatment or cold working.	Experience is required to interpret the test results.	Low cost; simple; portable units available; can be adapted to many comparative examinations.	Surface preparation may be necessary. Minimum thickness and width limitations exist. Assessment will be mainly qualitative.	1, 7, 8
Impact tests	Charpy and Izod tests are used to determine the susceptibility of material to brittle fracture. Provides a comparative and relative measure of fracture toughness of material by measuring impact energy, lateral expansion and fracture appearance. Drop-weight method is used to establish nil-ductility transition temperature.	A standard notched sample of metal is fractured under the impact energy of a dropping mass. The energy to cause fracture is a comparative measure of ductility and fracture toughness. Charpy V-notch beam and Izod notched cantilever tests are two common impact tests. Another common impact test is the Drop-weight test employed to establish nil-ductility transition temperature.	Moderate; can be performed by any qualified laboratory.	Relatively simple and inexpensive method to assess fracture toughness and resistance to shock loads. Fractured surface gives an indication of ductile or brittle nature of fracture.	Difficulties in obtaining nondestructive samples from the structure for testing. Test results are sensitive to composition and heat treatment of material and other operations and are affected by residual stresses.	1, 2, 9, 12

Modulus test (usually part of tension test)	To determine Young's Modulus, Tangent Modulus, and Chord Modulus of metallic materials.	Test specimen is loaded uniaxially and load and strain measured. The appropriate slope is calculated from the resultant stress-strain curve.	Moderate, can be performed by any qualified laboratory.	Gives fast and accurate results.	Samples must be removed from structure.	11
Tension test	To obtain yield strength, yield point, tensile strength, elongation and reduction of area. Sometimes a stress-strain or load elongation curve is also recorded during the test that can be useful in evaluating ductility and modulus of elasticity.	A standard metal specimen is subjected to axial tension and major mechanical properties are measured.	Low, can be performed by any qualified laboratory.	Gives fast and accurate results on mechanical properties of metals.	Samples must be removed from the structure for testing.	4, 5, 22

References for Table 3.7.6
1. ASTM A370. (1996). "Standard Test Methods and Definitions for Mechanical Testing of Steel Products," 1997 Annual Book of ASTM Standards, Vol.01.04, American Society for Testing and Materials, West Conshohocken, Pennsylvania, 164–209.
2. ASTM A673/A673M. (1995). "Standard Specification for Sampling Procedure for Impact Testing of Structural Steel," 1997 Annual Book of ASTM Standards, Vol 01.04, American Society for Testing and Materials, West Conshohocken, Pennsylvania, 329–332.
3. ASTM A751. (1995). "Standard Test Methods, Practices, and Terminology for Chemical Analysis of Steel Products," 1997 Annual Book of ASTM Standards, Vol. 01.04, American Society for Testing and Materials, West Conshohocken, Pennsylvania, 385–389.
4. ASTM E8. (1996a). "Standard Test Methods of Tension Testing of Metallic Materials," 1997 Annual Book of ASTM Standards, Vol. 03.01, American Society for Testing and Materials, West Conshohocken, Pennsylvania, 56–76.
5. ASTM E8M. (1996a). "Standard Test Methods for Tension Testing of Metallic Materials (Metric)," 1997 Annual Book of ASTM Standards, Vol. 03.01, American Society for Testing and Materials, West Conshohocken, Pennsylvania, 77–97.
6. ASTM E9. (1995). "Standard Test Methods of Compression Testing of Metallic Materials at Room Temperature," 1997 Annual Book of ASTM Standards, Vol. 03.01, American Society for Testing and Materials, West Conshohocken, Pennsylvania, 98–105.
7. ASTM E10. (1996). "Standard Test Method for Brinell Hardness of Metallic Materials," 1997 Annual Book of ASTM Standards, Vol. 03.01, American Society for Testing and Materials, West Conshohocken, Pennsylvania, 106–114.
8. ASTM E18. (1994). "Standard Test Methods for Rockwell Hardness and Rockwell Superficial Hardness of Metallic Materials," 1997 Annual Book of ASTM Standards, Vol. 03.01, American Society for Testing and Materials, West Conshohocken, Pennsylvania, 115–128.
9. ASTM E23. (1996). "Standard Test Methods for Notched Bar Impact Testing of Metallic Materials," 1997 Annual Book of ASTM Standards, Vol. 03.01, American Society for Testing Materials, West Conshohocken, Pennsylvania, 137–156.

References for Table 3.7.6 (*Continued*)

10. ASTM E30. (1989). "Standard Test Methods for Chemical Analysis of Steel, Cast Iron, Open-Hearth Iron, and Wrought Iron," 1994 Annual Book of ASTM Standards, Vol. 03.05, American Society for Testing and Materials, West Conshohocken, Pennsylvania, 39–50.
11. ASTM E111. (1996). "Standard Test Method for Young's Modulus, Tangent Modulus, and Chord Modulus," 1997 Annual Book of ASTM Standards, Vol. 03.01, American Society for Testing and Materials, West Conshohocken, Pennsylvania, 221–226.
12. ASTM E208. (1995a). "Standard Test Method for Conducting Drop-Weight Test to Determine Nil-Ductility Transition Temperature of Ferritic Steels," 1997 Annual Book of ASTM Standards, Vol.03.01, American Society for Testing and Materials, West Conshohocken, Pennsylvania, 294–305.
13. ASTM E399. (1997). "Standard Test Method for Plane-Strain Fracture Toughness of Metallic Materials," 1997 Annual Book of ASTM Standards, Vol. 03.01, American Society for Testing and Materials, West Conshohocken, Pennsylvania, 408–438.
14. ASTM E466. (1996). "Standard Practice for Conducting Force Controlled Amplitude Axial Fatigue Tests of Metallic Materials," 1997 Annual Book of ASTM Standards, Vol. 03.01, American Society for Testing and Materials, West Conshohocken, Pennsylvania, 466–470.
15. ASTM E561. (1994). "Standard Practice for R-Curve Determination," 1997 Annual Book of ASTM Standards, Vol. 03.01, American Society for Testing and Materials, West Conshohocken, Pennsylvania, 489–501.
16. ASTM E606. (1992). "Standard Practice for Strain Controlled Fatigue Testing," 1997 Annual Book of ASTM Standards, Vol. 03.01, American Society for Testing and Materials, West Conshohocken, Pennsylvania, 523–537.
17. ASTM E739. (1991). "Standard Practice for Statistical Analysis of Linear or Linearized Stress-Live (S-N) and Strain-Life (u-N) Fatigue Data," 1997 Annual Book of ASTM Standards, Vol. 03.01, American Society for Testing and Materials, West Conshohocken, Pennsylvania, 594–600.
18. ASTM E812. (1991). "Standard Test Method for Crack Strength of Slow-Bend Precracked Charpy Specimens of High-Strength Metallic Materials," 1997 Annual Book of ASTM Standards, Vol. 03.01, American Society for Testing and Materials, West Conshohocken, Pennsylvania, 623–626.
19. ASTM E813. (1989). "Standard Test Method for J_{Ic}, A Measure of Fracture Toughness," 1997 Annual Book of ASTM Standards, Vol. 03.01, American Society for Testing and Materials, West Conshohocken, Pennsylvania, 627–641.
20. ASTM E1049. (1990). "Standard Practices for Cycle Counting in Fatigue Analysis," 1997 Annual Book of ASTM Standards, Vol. 03.01, American Society for Testing and Materials, West Conshohocken, Pennsylvania, 707–715.
21. ASTM F606. (1995b). "Standard Test Method for Determining the Mechanical Properties of Externally and Internally Threaded Fasteners, Washers, and Rivets," 1997 Annual Book of ASTM Standards, Vol. 15.08, American Society for Testing and Materials, West Conshohocken, Pennsylvania, 188–199.
22. American Society for Metals. (1985). "Metals Handbook: Mechanical Testing, Ninth Edition," Volume 8, AMS, Metals Park, Ohio.

SEI/ASCE 11-99

TABLE 3.7.7. NDT Methods for Determining Chemical and Physical Properties of Steel

Test Method	Purpose of Test	Principle of Operation	User Expertise	Advantages	Limitations	References
Hardness tests	Using a hardness tester (sometimes portable), a hard steel ball or diamond shape penetrator is forced into the material using standard procedures. By measuring the geometry of the indentation mark left on the material, hardness number is established. Several hardness numbers are defined and used in industry. Brinell and Rockwell hardness numbers are used frequently to define hardness of metals.	To evaluate resistance to permanent or plastic deformations. To obtain quick but approximate value of tensile strength. Also used to evaluate quality level and uniformity of heat treatment or cold working.	Experience is required to interpret the test results.	Low cost; simple; portable units available; can be adapted to many comparative examinations. A permanent record is obtained.	Surface preparation may be necessary. Minimum thickness and width limitations exist. Assessment will be mainly qualitative. Access to structure and means of holding tester may be limited.	1

References for Table 3.7.7.
1. ASTM E110. (1992). "Standard Test Method for Indentation Hardness of Metallic Materials by Portable Hardness Testers," 1997 Annual Book of ASTM Standards, Vol. 03.01, American Society for Testing and Materials, West Conshohocken, Pennsylvania, 219–220.

TABLE 3.7.8. NDT Methods for Determining Physical Conditions of Steel

Test Method	Purpose of Test	Principle of Operation	User Expertise	Advantages	Limitations	References
Acoustic emission	To detect locations of high stress concentration or plastic yielding while structure is stressed. Can be used in proof loading to monitor and detect areas of high stress intensity.	The structure to be investigated is subjected to a mechanical or external stimulus that produces release of strain energy. The rate of release of strain energy will be relatively high at crack tips and at local yield points. The release of strain energy generates high-frequency acoustic stress waves that can be detected by acoustic sensors attached to the structure.	Expertise required to plan, conduct and interpret test results.	Nondestructive only if the required loading does not cause damage. Can be used to detect stress raisers and local yield areas. Can be used in long term monitoring of structure under loading.	Expensive; requires complex equipment. It can only detect active flaws that release strain energy under applied stress. Cannot detect static flaws that are not experiencing yielding or fracture slips. Is irreversible and by unloading and reloading the structure to the same stress level no emission will be produced.	4, 11
Bolt tightness	To establish effectiveness of friction of bolts.	Torque wrenches (calibrated in laboratory) are used to measure torque which is converted to bolt tension by use of tables.	Moderate expertise required.	Nondestructive, easy and inexpensive.	Not too reliable. Difficult to calibrate wrenches and to relate to tension.	17, 18
Electromagnetic (Eddy-current)	A means of sorting metals to detect surface discontinuities, dimensions, cracks, seams, variations in alloy composition or heat treatment. Used to evaluate condition of wire tubing; for thickness gauging; for electrically conductive or magnetically permeable metals.	Is based on change of impedance of a probe coil placed in contact with metal. The impedance of coil probe is dependent on material properties of metal specimen and its composition.	Experience is required to conduct the test and interpret the test results.	Nondestructive, moderate cost; readily automated; portable; permanent records available if needed; can be adapted to many comparative analyses.	Shallow penetration; reference standards often are necessary; no absolute measurements; only qualitative measurements. Laminar cracks not open to surface and crack planes perpendicular to the axis of eddy current cylinder may escape detection.	7, 11

Liquid penetrant	To detect discontinuities open to a surface. Used to detect small cracks, lamination, weld slag, incomplete fusion in welds that are open to surface, poor bond, gauges, porosity, laps, and fabricating discontinuities.	Liquid penetrant is applied to the surface. The liquid will penetrate discontinuities open to surface. The surface is cleaned and a developer is applied. Due to capillary action the liquid rises to the surface and marks the flaw or crack on the developer. Two types of liquid penetrants are used; dye penetrant that can be detected visually and fluorescent penetrant that can be detected in darkness or when exposed to ultraviolet light.	Experience required for application of liquid and developer. Expertise required to interpret the results.	Nondestructive; simple; inexpensive; portable; allows inspection of complex shapes in one single operation. Applicable to magnetic or nonmagnetic material. Can be used to verify results of magnetic particle testing. An indication of depth of crack can be obtained by examining the width of bleedout.	Only defects open to surface can be detected. Irrelevant indications can occur due to surface conditions. Temperature of sample, penetrant drain time, emulsifier soak and drain time, drying temperature and developing powder dwell time must be controlled carefully.	3, 11, 12, 13
Magnetic particle	To locate surface and slightly below surface discontinuities such as cracks, inclusions, lamination, porosities, voids, and laps.	The area to be inspected is magnetized. Then the area is covered by fine powder of magnetic material. A visible pattern of discontinuity will develop due to stronger attraction at those points.	Low to moderate to apply. Experience needed to interpret results.	Nondestructive; simpler than radiographic inspection; relatively inexpensive. Senses surface flaws and subsurface flaws down to about 0.25 in. (6 mm) below the surface. Cracks filled with foreign objects can be detected.	Applicable only to magnetic metals. It is messy. Careful surface preparation is required. Irrelevant indications can occur. Depends on operator's ability to interpret the results. Defects and discontinuities parallel to magnetic field may not give pattern. Applications along two perpendicular axes may be necessary. Demagnetization after test may be needed.	5, 11, 12, 13

TABLE 3.7.8. NDT Methods for Determining Physical Conditions of Steel (*Continued*)

Test Method	Purpose of Test	Principle of Operation	User Expertise	Advantages	Limitations	References
Radiography	To measure microscopic voids, porosity, inclusions and cracks. In welding it can be used to detect blow holes, nonmetallic inclusions, incomplete penetration, and undercutting in addition to detecting above mentioned defects.	Is based on the principle of radiography. The x-ray or gamma ray introduced to sample defects at voids or low density areas. As a result dark spots will appear at these areas on radiograph films.	Experience required to operate. Expertise required to interpret radiograph films.	Nondestructive; detects both internal and external flaws; portable; provides a permanent record on X-ray film.	High cost; heavy; health hazard; not sensitive to defects less than 2% of the total thickness of the specimen. Complex shapes are difficult to analyze. Very small cracks or cracks not essentially parallel to the radiation beam such as lamination cracks cannot be detected.	1, 8, 11, 12, 13
Residual stress by hole-drilling	To measure principal stresses on the surface of material.	A hole is drilled in the area where residual stresses are to be measured. The stresses in the vicinity of the hole are relaxed. By attaching three element rosette strain gauges, the value of relaxed stresses is measured. The relieved surface strain is related to principal stresses that were present before drilling the hole.	Moderate to conduct the test.	Semidestructive with only damage due to small hole and cemented strain gauges. A relatively simple method to measure stresses in existing structure.	Applicable only to easily accessible surfaces where strain gauges can properly be mounted; interpretation of residual versus load induced stresses is difficult.	6

Ultrasonic	To detect cracks, voids, porosities, laps, poor brazing and bounding, and segregated inclusions including flaws too small to be detected by other methods.	Ultrasonic waves are applied to the surface of metal. The waves will propagate in the metal and will be reflected by any discontinuity or change in density. The reflected echo is analyzed to locate internal flaws and cracks.	High; for development of test procedures, operation and interpretation of results.	Nondestructive; portable; locates very small surface and subsurface discontinuities; instant results; accurate measure of thickness. Because of good penetration can be used to inspect thick objects. Extremely small flaws can be detected.	Sensitivity is reduced by rough surface; odd shape samples are hard to analyze; requires skilled operator. Depends on orientation of defects and operator's ability to interpret results. Must couple transducer to surface of specimen carefully. Permanent record is not obtained. Referenced standards may be needed for calibration of equipment and for interpretation of results.	2, 10, 11, 12, 13, 15, 16, 19, 20
Visual	To evaluate surface condition. To check dimensional accuracy. To detect corrosion, cracks, voids, fabricating discontinuities, weld undercuts and overlaps, pits and other surface irregularities. To evaluate condition of connections and connectors, missing or loose bolts, slag inclusions, unwelded and poor weld appearances.	A visual examination of surface condition of metals can be performed by naked eye or by the use of special devices such as magnifiers, borescope, fiberoptic, flashlight, weld gauge, and panoramic camera. Magnifiers can be used to detect small flaws and cracks.	Experience and expertise is required for thorough inspection and detection of flaws.	Nondestructive; simple; inexpensive; portable; large amount of data on surface condition and overall appearance of members can be obtained. Can be used to locate problem areas. Hidden surfaces can be inspected using special devices. Should be used in any evaluation even if other more powerful methods of testing are used.	Basically qualitative information is obtained. Little quantitative measurements can be made. Only surface condition can be evaluated. Some permanent records can be obtained by photography and recording the condition.	9, 14

SEI/ASCE 11-99

69

References for Table 3.7.8

1. ASTM E94. (1993). "Standard Guide for Radiographic Testing," 1997 Annual Book of ASTM Standards, Vol. 03.03, American Society for Testing and Materials, West Conshohocken, Pennsylvania, 1–11.
2. ASTM E164. (1994a). "Standard Practice for Ultrasonic Contact Examination of Weldments," 1997 Annual Book of ASTM Standards, Vol. 03.03, American Society for Testing and Materials, West Conshohocken, Pennsylvania, 36–55.
3. ASTM E165. (1995). "Standard Practice for Liquid Penetrant Inspection Method," 1997 ASTM Annual Book of ASTM Standards, Vol. 03.03, American Society for Testing and Materials, West Conshohocken, Pennsylvania, 56–75.
4. ASTM E569. (1991). "Standard Practice for Acoustic Emission Monitoring of Structures During Controlled Stimulation," 1997 Annual Book of ASTM Standards, Vol. 03.03, American Society for Testing and Materials, West Conshohocken, Pennsylvania, 228–231.
5. ASTM E709. (1995). "Standard Guide for Magnetic Particle Examination," 1997 Annual Book of ASTM Standards, Vol. 03.03, American Society for Testing and Materials, West Conshohocken, Pennsylvania, 279–310.
6. ASTM E837. (1995). "Standard Test Method for Determining Residual Stresses by the Hole-Drilling Strain-Gage Method," 1997 Annual Book of ASTM Standards, Vol. 03.01, American Society for Testing and Materials, West Conshohocken, Pennsylvania, 642–648.
7. ASTM E1004. (1991). "Standard Test Method for Electromagnetic (Eddy-Current) Measurements of Electrical Conductivity," 1997 Annual Book of ASTM Standards, Vol. 03.03, American Society for Testing and Materials, West Conshohocken, Pennsylvania, 429–432.
8. ASTM E1032. (1995). "Standard Test Method for Radiographic Examination of Weldments," 1997 Annual Book of ASTM Standards, Vol. 03.03, American Society for Testing and Materials, West Conshohocken, Pennsylvania, 449–453.
9. ASTM G46. (1994). "Standard Guide for Examination and Evaluation of Pitting Corrosion," 1997 Annual Book of ASTM Standards, Vol. 03.02, American Society for Testing and Materials, West Conshohocken, Pennsylvania, 169–175.
10. Al-Nassar, Y., Datta, S. K., and Shah, A. H. (1990). "Lamb Wave Scattering by a Surface-breaking Crack in a Welded Plate," *Proc., Non-Destructive Evaluation of Civil Structures and Materials*, edited by B. A. Suprenant, S. Sture, J. L. Noland, and M. P. Schuller, University of Colorado, Boulder, Colorado, October, 297–314.
11. American Society for Metals. (1976). "Metals Handbook: Nondestructive Inspection and Quality Control," ASM, Metals Park, Ohio.
12. American Welding Society. (1986). "Guide for the Nondestructive Inspection of Welds," Designation ANSI/AWS B.1.10-86, American Welding Society, Inc., Miami, Florida.
13. American Welding Society. (1990). "Structural Welding Code-Steel," Designation ANSI/AWS D1.1-90, American Welding Society, Inc., Miami, Florida.
14. Ferris, H. W. (1953). "Historical Record, Dimensions and Properties, Rolled Shapes, Steel and Wrought Iron Beams and Columns, as Rolled in U.S.A., Period 1873–1952 with Source Noted," American Institute of Steel Construction, Chicago, Illinois.
15. Førli, O. (1990). "Non-Destructive Evaluation of Steel Structures—Techniques and Reliability," *Proc., Non-Destructive Evaluation of Civil Structures and Materials*, edited by B. A. Suprenant, S. Sture, J. L. Noland, and M. P. Schuller, University of Colorado, Boulder, Colorado, October.
16. Jessop, T. J., Mudge, P. J., and Harrison, J. D. (1981). "Ultrasonic Measurement of Weld Flaw Size," National Cooperative Highway Research Program Report 242, Transportation Research Board, National Research Council, Washington, D.C., December.
17. Research Council on Structural Connections. (1985). "Allowable Stress Design Specification for Structural Joints Using ASTM A325 or A490 Bolts," Manual of Steel Construction, Allowable Stress Design, American Institute of Steel Construction, Chicago, 1989, Illinois, 5-263 to 5-307.
18. Research Council on Structural Connections. (1988). "Load and Resistance Factor Design Specification for Structural Joints Using ASTM A325 or A490 Bolts," Manual of Steel Construction, Load and Resistance Factor Design, American Institute of Steel Construction, 1994, Chicago, Illinois, 6-371 to 6-422.
19. Sakamoto, K., Fukazawa, M., Hamano, M., and Tajima, J. (1985). "Estimation of Fatigue Crack Growth by Ultrasonic Imaging Method," *Proc., Structural Engineering/Earthquake Engineering*, Japan Society of Civil Engineers, Vol. 2, No. 2, October, 455s–465s.
20. U.S. Department of Transportation, Federal Highway Administration. (1968). "Ultrasonic Testing Inspection," Washington, D.C.

TABLE 3.7.9. Test Methods for Determining Physical Properties of Masonry Units and Masonry Assemblages on Masonry Specimens

Property	Purpose of Test	Test Method	User Expertise	Advantages	Limitations	References
Absorption	Determination of the percentage of absorption by mass of brick, tile and concrete masonry units and of natural building stone.	Determination of both dry and saturated mass.	Standard laboratory test. Moderate expertise required.	Easy test, low cost.	Specimens must be taken from structure for testing. Results are affected by mortar clogging pores.	3, 4, 5, 6, 7, 8, 9, 10, 15, 16, 18, 19, 20, 23, 24, 25, 26, 27, 47
Bulk specific gravity	Determination of bulk specific gravity of natural building stone.	Determination of weights of dried specimen, soaked and surface dried specimen and of soaked specimen in water.	Standard laboratory test. Moderate expertise required.	Easy test, low cost.	Specimens must be taken from structure for testing.	10, 22, 23, 24, 25
Compressive strength	Determination of compressive strength of brick, tile and concrete masonry units and natural building stone.	Specimens of masonry units or stone are capped and tested in a testing machine using standard procedures.	For test on specimens, standard laboratory tests. Moderate expertise required.	Easy tests, low cost.	Test specimens must be taken from structure for testing.	3, 4, 5, 6, 7, 8, 9, 14, 15, 16, 17, 18, 19, 23, 24, 25, 26, 27, 30, 38
Compressive strength (in situ)	Compressive strength of masonry, in situ	In situ tests involve installation of flatjack device in saw-cut mortar joints.	For in situ tests, moderate expertise required.	In situ test possible.	For in situ tests, estimated stress restricted to width of flatjack.	33
Concentrated load	Determination of the concentrated loads that can be supported by masonry assemblages.	Determination of average compressive stress on unreinforced solid-unit masonry by loading.	Specialized expertise required.	In situ test possible.	Special jigs required. Expensive.	37
Deformability	Determination of the deformability of masonry assemblages.	In situ test involving installation of flatjack device in saw-cut mortar joints.	Low degree of expertise required.	Easy test.	Relatively nondestructive.	34
Durability (See also freezing and thawing)	Determination of durability of brick masonry as measured by compressive strength, absorption and saturation coefficient.	Standard tests for compressive strength, absorption and saturation coefficient.	Standard laboratory tests. Moderate expertise required.	Easy tests, low cost.	Specimens must be taken from structure for testing. Results may be misleading. Freezing and thaw test may provide better results.	6, 7, 8, 19, 27

SEI/ASCE 11-99

TABLE 3.7.9. Test Methods for Determining Physical Properties of Masonry Units and Masonry Assemblages on Masonry Specimens *(Continued)*

Property	Purpose of Test	Test Method	User Expertise	Advantages	Limitations	References
Efflorescence	Determination of leaching of soluble salts in brick masonry units.	Specimens are immersed in distilled water then oven dried, inspected and rated.	Standard laboratory test. Low degree of expertise required.	Easy test, low cost.	False results possible since units are contaminated in the wall.	19, 27
Flexure strength	Determination of flexural strength of building stone.	Simple beam testing using quarter point loading.	Standard laboratory test. Moderate expertise required.	Useful because it indicates differences in flexural strength between different building stones.	Fairly long sample (ten times depth) required to be taken from structure.	24, 29
Flexural tensile strength	Determination of tensile strength that can be supported by masonry assemblages.	Determination of average flexural tensile strength of unreinforced solid-unit masonry by loading.	Specialized expertise required.	In situ test possible.	Special jigs required. Expensive.	32, 37, 41
Freezing and thawing	Determination of resistance of brick masonry units to rapidly repeated cycles of freezing and thawing.	Test specimens under controlled moisture conditions are subjected to standard cycles of freezing and thawing. Specimens are measured for mass before and after tests and percent changes recorded.	Standard laboratory tests. Moderate expertise required.	Provides useful information.	Specimens must be taken from structure for testing. Requires 10 weeks to complete test.	6, 7, 8, 19, 27, 29, 35
Initial rate of absorption	Determination of initial rate of absorption of brick masonry units and slate.	Specimens are dried, set in water for 1 min, and mass determined. Gain in mass is recorded.	Standard laboratory test. Moderate expertise required.	Important information to determine loss of mixing water in mortar and hence water tightness of joints.	Specimens must be taken from structure for testing.	6, 7, 8, 13, 19, 26, 27
Modulus of rupture and secant modulus of rupture	Determination of modulus of rupture of natural building stone and of brick masonry units and modulus of rupture and secant modulus of rupture of masonry assemblage.	Specimens are tested as beam with midspan loading and tested to failure. Modulus of rupture calculated from breaking load.	Standard laboratory test. Moderate expertise required.	Test indicates differences between various specimens. ASTM E72 and E519 can be adapted to in situ tests.	Specimens must be taken from structure for testing.	7, 8, 11, 12, 20, 23, 24, 25, 26, 37, 42

SEI/ASCE 11-99

Saturation	Determination of saturation coefficient of brick masonry units.	Same procedure as for determining absorption except saturated mass after 24 h and 5 h boiling are used in calculation.	Standard laboratory test. Low degree of expertise required.	Saturation coefficient used as measure of durability.	Specimens must be taken from structure for testing. Results may be misleading.	6, 7, 19, 27
Shear strength	Determination of shear strength that can be supported by masonry assemblages.	Determination of average shear strength in unreinforced solid-unit masonry by loading.	Specialized expertise required.	ASTM E72 and E519 can be adapted to in situ test. UBC 21-6 is an in situ test.	Special jigs required. Expensive. UBC 21.6 has limited use.	30, 37, 42, 46
Splitting tensile strength	Determination of splitting tensile strength of masonry units. Can also be used to detect the presence of defects.	Compressive load is applied to the unit by means of bearing rods resulting in a tensile stress distributed over the height of the unit.	Specialized expertise required.	Laboratory method to determine physical properties of masonry.	Precision of test method not very good. Currently, seldom used to determine physical properties.	31
Strength of anchors	Determination of static, tensile and shear strength of post-installed and cast-in-place anchorage systems in structural masonry.	Loading plates are located at anchors to be tested and appropriate testing equipment and testing system used to determine strength of anchor.	Special jigs must be constructed to mount on to anchors in structure. Specialized expertise required.	In situ strength values of anchorage determined.	While specimens are not removed from structure, tests are destructive.	39, 43
Tensile load	Determination of tensile load that can be supported by masonry assemblages.	Determination of average tensile load of unreinforced solid unit masonry by loading.	Specialized expertise required.	ASTM E72 can be adapted to in situ test.	Special jigs required. Expensive.	30, 37, 41
Transverse load	Determination of transverse load that can be supported by masonry assemblages.	Determination of average transverse load on unreinforced solid unit masonry by loading.	Specialized expertise required.	ASTM E72 can be adapted to in situ test.	Special jigs required. Expensive.	30, 37, 41
Water penetration	Determination of resistance to water penetration of masonry assemblages.	Test chambers are placed against specimens with controlled temperature and humidity conditions. Water is applied with increased air pressure. Dampness on back of specimen observed.	Standard laboratory test. Specialized expertise required.	Provides useful information on quality of material coatings, leakage and workmanship.	Water penetration is significantly affected by air pressure in test chamber. ASTM E514 intended for laboratory use.	40, 45

References for Table 3.7.9

1. ACI 530/ASCE 5/TMS 402. (1995). "Building Code for Masonry Structures," American Society of Civil Engineers, Reston, Virginia.
2. ACI 530.1/ASCE 6/TMS 602. (1995). "Specifications for Masonry Structures," American Society of Civil Engineers, Reston, Virginia.
3. ASTM C34. (1996). "Standard Specification for Structural Clay Load-Bearing Tile," 1997 Annual Book of ASTM Standards, Vol. 4.05, American Society for Testing and Materials, West Conshohocken, Pennsylvania, 22–24.
4. ASTM C55. (1997). "Standard Specification for Concrete Building Brick," 1997 Annual Book of ASTM Standards, Vol. 4.05, American Society for Testing and Materials, West Conshohocken, Pennsylvania, 29–31.
5. ASTM C56. (1996). "Standard Specification for Structural Clay Non-Load-Bearing Tile," 1997 Annual Book of ASTM Standards, Vol. 4.05, American Society for Testing and Material, West Conshohocken, Pennsylvania, 32–33.
6. ASTM C62. (1997). "Standard Specification for Building Brick (Solid Masonry Units Made from Clay or Shale)," 1997 Annual Book of ASTM Standards, Vol. 04.05, American Society for Testing and Materials, West Conshohocken, Pennsylvania, 34–37.
7. ASTM C67. (1997). "Standard Test Methods of Sampling and Testing Brick and Structural Clay Tile," 1997 Annual Book of ASTM Standards, Vol. 04.05, American Society for Testing and Materials, West Conshohocken, Pennsylvania, 38–47.
8. ASTM C73. (1997). "Standard Specification for Calcium Silicate Face Brick (Sand-Lime Brick)," 1997 Annual Book of ASTM Standards, Vol. 4.05, American Society for Testing and Materials, West Conshohocken, Pennsylvania, 48–49.
9. ASTM C90. (1997). "Standard Specification for Load-Bearing Concrete Masonry Units," 1997 Annual Book of ASTM Standards, Vol. 4.05, American Society for Testing and Materials, West Conshohocken, Pennsylvania, 71–75.
10. ASTM C97. (1996). "Standard Test Methods for Absorption and Bulk Specific Gravity of Dimension Stone," 1997 Annual Book of ASTM Standards, Vol. 4.07, American Society for Testing and Materials, West Conshohocken, Pennsylvania, 1–2.
11. ASTM C99. (1993). "Standard Test Method for Modulus of Rupture of Dimension Stone," 1997 Annual Book of ASTM Standards, Vol. 4.07, American Society for Testing and Materials, West Conshohocken, Pennsylvania, 3–5.
12. ASTM C120. (1994). "Standard Test Method for Flexural Testing of Slate (Modulus of Rupture, Modulus of Elasticity)," 1997 Annual Book of ASTM Standards, Vol. 4.07, American Society for Testing and Materials, West Conshohocken, Pennsylvania, 10–11.
13. ASTM C121. (1994). "Standard Test Method for Water Absorption of Slate," 1997 Annual Book of ASTM Standards, Vol. 4.07, American Society for Testing and Materials, West Conshohocken, Pennsylvania, 12–13.
14. ASTM C126. (1996). "Standard Specification for Ceramic Glazed Structural Clay Facing Tile, Facing Brick, and Solid Masonry Units," 1997 Annual Book of ASTM Standards, Vol. 4.05, American Society for Testing and Materials, West Conshohocken, Pennsylvania, 82–86.
15. ASTM C129. (1997). "Standard Specification for Non-Load-Bearing Concrete Masonry Units," 1997 Annual Book of ASTM Standards, Vol. 4.05, American Society for Testing and Materials, West Conshohocken, Pennsylvania, 87–89.
16. ASTM C140. (1996b). "Standard Test Methods of Sampling and Testing Concrete Masonry Units," 1997 Annual Book of ASTM Standards, Vol. 4.05, American Society for Testing and Materials, West Conshohocken, Pennsylvania, 92–99.
17. ASTM C170. (1994). "Standard Test Method for Compressive Strength of Dimension Stone," 1997 Annual Book of ASTM Standards, Vol. 04.07, American Society for Testing and Materials, West Conshohocken, Pennsylvania, 14–16.
18. ASTM C212. (1996). "Standard Designation for Structural Clay Facing Tile," 1997 Annual Book of ASTM Standards, Vol. 4.05, American Society for Testing and Materials, West Conshohocken, Pennsylvania, 107–111.
19. ASTM C216. (1997). "Standard Specification for Facing Brick (Solid Masonry Units Made from Clay or Shale)," 1997 Annual Book of ASTM Standards, Vol. 04.05, American Society for Testing and Materials, West Conshohocken, Pennsylvania, 112–116.
20. ASTM C279. (1995). "Standard Specification for Chemical-Resistant Masonry Units," 1997 Annual Book of ASTM Standards, Vol. 4.05, American Society for Testing and Materials, West Conshohocken, Pennsylvania, 147–148.
21. ASTM C426. (1996a). "Standard Test Method for Linear Drying Shrinkage of Concrete Masonry Units," 1997 Annual Book of ASTM Standards, Vol. 4.05, American Society for Testing and Materials, West Conshohocken, Pennsylvania, 244–248.
22. ASTM C503. (1996). "Standard Specification for Marble Building Stone (Exterior)," 1997 Annual Book of ASTM Standards, Vol 4.07, American Society for Testing and Materials, West Conshohocken, Pennsylvania, 25–26.
23. ASTM C568. (1996). "Standard Specification for Limestone Building Stone," 1997 Annual Book of ASTM Standards, Vol. 4.07, American Society for Testing Materials, West Conshohocken, Pennsylvania, 41–42.

24. ASTM C615. (1996). "Standard Specification for Granite Building Stone," 1997 Annual Book of ASTM Standards, Vol 4.07, American Society for Testing and Materials, West Conshohocken, Pennsylvania, 47–48.
25. ASTM C616. (1996). "Standard Specification for Quartz-Based Dimension Stone," 1997 Annual Book of ASTM Standards, Vol. 4.07, American Society for Testing and Materials, West Conshohocken, Pennsylvania, 49–50.
26. ASTM C629. (1996). "Standard Specification for Slate Building Stone," 1997 Annual Book of ASTM Standards, Vol. 4.07, American Society for Testing and Materials, West Conshohocken, Pennsylvania, 51–52.
27. ASTM C652. (1997). "Standard Specification for Hollow Brick (Hollow Masonry Units Made from Clay or Shale)," 1997 Annual Book of ASTM Standards, Vol. 4.05, American Society for Testing and Materials, West Conshohocken, Pennsylvania, 378–382.
28. ASTM C744. (1997). "Standard Specification for Prefaced Concrete and Calcium Silicate Masonry Units," 1997 Annual Book of ASTM Standards, Vol. 04.07, American Society for Testing and Materials, West Conshohocken, Pennsylvania, 428–430.
29. ASTM C880. (1993). "Standard Test Method for Flexural Strength of Dimension Stone," 1997 Annual Book of ASTM Standards, Vol. 4.05, American Society for Testing and Materials, West Conshohocken, Pennsylvania, 146–147.
30. ASTM C901. (1993A). "Standard Specification for Prefabricated Masonry Panels," 1997 Annual Book of ASTM Standards, Vol. 4.05, American Society for Testing and Materials, West Conshohocken, Pennsylvania, 565–567.
31. ASTM C1006. (1996). "Standard Test Method for Splitting Tensile Strength of Masonry Units," 1997 Annual Book of ASTM Standards, Vol. 4.05, American Society for Testing and Materials, West Conshohocken, Pennsylvania, 658–660.
32. ASTM C1072. (1994). "Standard Test Method for Measurement of Masonry Flexural Bond Strength," 1997 Annual Book of ASTM Standards, Vol. 4.05, American Society for Testing and Materials, West Conshohocken, Pennsylvania, 666–671.
33. ASTM C1196. (1992). "Standard Test Method for In Situ Compressive Stress Within Solid Unit Masonry Estimated Using Flatjack Measurements," 1997 Annual Book of ASTM Standards, Vol. 4.05, American Society for Testing and Materials, West Conshohocken, Pennsylvania, 780–784.
34. ASTM C1197. (1992). "Standard Test Method for In Situ Measurement of Masonry Deformability Properties Using the Flatjack Method," 1995 Annual Book of ASTM Standards, Vol. 4.05, American Society for Testing and Materials, West Conshohocken, Pennsylvania, 785–789.
35. ASTM C1262. (1996). "Standard Test Method for Evaluating the Freeze-Thaw Durability of Manufactured Concrete Masonry and Related Concrete Units," 1997 Annual Book of ASTM Standards, Vol. 4.05, American Society for Testing and Materials, West Conshohocken, Pennsylvania, 836–839.
36. ASTM C1324. (1996). "Standard Test Method for Examination and Analysis of Hardened Masonry Mortar," 1997 Annual Book of ASTM Standards, Vol. 4.02, American Society for Testing and Materials, West Conshohocken, Pennsylvania, 871–876.
37. ASTM E72. (1980). "Standard Method of Conducting Strength Tests of Panels for Building Construction," 1994 Annual Book of ASTM Standards, Vol. 4.07, American Society for Testing and Materials, West Conshohocken, Pennsylvania, 369–379.
38. ASTM E447. (1997). "Standard Test Methods for Compressive Strength of Laboratory Constructed Masonry Prisms," 1997 Annual Book of ASTM Standards, Vol. 4.05, American Society for Testing and Materials, West Conshohocken, Pennsylvania, 1014–1016.
39. ASTM E488. (1990). "Standard Test Method for Strength of Anchors on Concrete and Masonry Units," 1994 Annual Book of ASTM Standards, Vol. 04.07, American Society for Testing and Materials, West Conshohocken, Pennsylvania, 509–516.
40. ASTM E514. (1990). "Standard Test Method for Water Penetration and Leakage Through Masonry," 1995 Annual Book of ASTM Standards, Vol. 04.07, American Society for Testing and Materials, West Conshohocken, Pennsylvania, 903–906.
41. ASTM E518. (1993). "Standard Test Methods for Flexure Bond Strength of Masonry," 1997 Annual Book of ASTM Standards, Vol. 4.05, American Society for Testing and Materials, West Conshohocken, Pennsylvania, 1021–1024.
42. ASTM E519. (1993). "Standard Test Method for Diagonal Tension (Shear) in Masonry Assemblages," 1997 Annual Book of ASTM Standards, Vol. 4.05, American Society for Testing and Materials, West Conshohocken, Pennsylvania, 1025–1028.
43. ASTM E754. (1994). "Standard Test Method for Pullout Resistance of Ties and Anchors Embedded in Masonry Mortar Joints," 1994 Annual Book of ASTM Standards, Vol. 04.07, American Society for Testing and Materials, West Conshohocken, Pennsylvania, 701–708.
44. Atkinson, R. (1993). "Evaluating the Structural Condition of Existing Masonry," Concrete Repair Bulletin, March/April, 14–16.
45. Krogstad, N. V. (1990). "Masonry Wall Drainage Test—A Proposal Method for Field Evaluation of Masonry Cavity Walls for Resistance to Water Leakage," Masonry: Components to Assemblages, ASTM STP 1063, edited by J. H. Matthys, American Society for Testing and Materials, West Conshohocken, Pennsylvania, 394–402.
46. UBC 21-6. (1994). "In-place Masonry Shear Tests," Uniform Building Code, Section 21-6, International Conference of Building Officials, Whittier, California.
47. Yorkdale, A. H. (1982). "Initial Rate of Absorption and Mortar Bond," Masonry: Materials, Properties and Performance, ASTM STP 778, edited by J. G. Borchelt, American Society for Testing and Materials, West Conshohocken, Pennsylvania, 91–98.

TABLE 3.7.10. NDT Methods for Determining Properties and Physical Conditions of Masonry Units and Masonry Assemblages

Test Method	Purpose of Test	Principle of Operation	User Expertise	Advantages	Limitations	References
Acoustic impact (Hammer test)	Detect delaminations or disbonds in composite systems; detect voids and cracks in materials, e.g., hammer technique to detect defective masonry units.	Surface of object is struck with a hammer (usually metallic). The frequency and damping characteristics of the "ringing" can indicate the presence of defects.	Low level of expertise required to use, but experience needed for interpreting results.	Portable; easy to perform test; electronic measuring equipment not needed for qualitative results.	Geometry and mass of test object influences results; poor discrimination; reference standards required for electronic testing.	17, 35, 37
Electrical resistance probe	Measure of moisture migration patterns in masonry walls.	Electrical resistance between two probes inserted into test component is measured. The resistance decreases with increased moisture content.	Low.	Equipment is inexpensive, simple to operate, and many measurements can be rapidly made.	Not reliable at high moisture content; precise results are not usually obtained.	Table 3.7.3, 17
Flatjack	Determination of in situ stresses and deformability.	A portion of a mortar joint is removed, and a flatjack installed in the slot. Deformations are measured before the slot is cut and the jack pressure required to restore deformation.	Moderate expertise required.	Can determine stress in wall in cases where loadings or displacements are unknown or difficult to quantify. High reliability.	Numerous measurements are required to obtain a full picture of deformation and stress. Usually only one face of masonry wall is accessible.	9, 15, 24, 25, 29, 35
Fiber optics (probe holes with fiberscope)	Check condition of materials in cavity, such as thermal insulation, pipes, and electrical wiring installed in wall cavities; check for unfilled cores in reinforced masonry construction; check for voids along grouted stressed tendons.	Bundle of flexible, optical fibers with lens and illuminating systems is inserted into small bore holes thus enabling view of interior of cavities.	Low.	Direct visual inspection of otherwise inaccessible parts is possible.	Probe holes usually must be drilled; probe holes must connect to a cavity.	Table 3.7.3, 35

SEI/ASCE 11-99

Gamma radiography	Locates internal cracks, voids and variations in density and composition of materials. Locating internal parts in a structural component, e.g., reinforcing steel in masonry.	Gamma radiation attenuates when passing through a structural component. Extent of attenuation controlled by density, and thickness of materials of the structural component. Photographic film record usually made and analyzed.	High level of expertise.	Portable and relatively inexpensive compared to X-ray radiography; internal defects can be detected; applicable to a variety of materials.	Radiation intensity cannot be adjusted; long exposure times may be required; dangerous radiation; two opposite surfaces of component must be accessible. A special license is required to obtain gamma isotope sources.	3, 7, 10, 17, 19, 25, 35, 37
Physical measurements	Physical documentation of observed or suspected conditions observed visually.	Physical measurement of alignment, crack size and location, differential settlement of structure, deflections, displacement and distortion by means of surveying tools and instruments.	Moderate level of expertise required in use of surveying equipment. Expertise needed for interpretation of results.	Easy to perform. No special equipment. Low cost.	Physical access required. This may require scaffolding and could be costly.	12
Ultrasonic pulse velocity	Internal discontinuities can be located and their size estimated.	Operates on principle that vibrational wave propagation is affected by quality of masonry; pulse waves are induced in materials and those reflected back are detected; both the transmitting and receiving transducers usually are contained in the same probe.	High level of expertise required to interpret results. Low level of expertise required to make measurements.	Excellent for rapid surveys of large areas.	Good coupling between transducer and test substrates critical; interpretation of results can be difficult; density and amount of aggregate may affect results; calibration standards required.	1, 7, 17, 18, 20, 22, 25, 26, 27, 29, 31, 35, 37, 38
Visual examination	Evaluation of surface condition of masonry; determining deficiencies in joints; determining differential movement in joints; determining warping, bulging or sagging of components. Visual examination includes sampling of units.	Visual examination with or without optical aids, photographic records, or other low cost tools; differential movement determined over long periods of time with use of crack movement monitoring devices.	Experience required in order to determine what to look for, what measurements to take, and what follow-up testing to specify.	Generally low cost; rapid results except for surveying method.	Trained evaluator required; primary evaluation confined to surface of structure.	4, 5, 12, 14, 32, 33, 34, 36, 39

77

TABLE 3.7.10. NDT Methods for Determining Properties and Physical Conditions of Masonry Units and Masonry Assemblages (*Continued*)

Test Method	Purpose of Test	Principle of Operation	User Expertise	Advantages	Limitations	References
Water penetration	Determine resistance to leakage and water penetration. Test is basically a laboratory test.	Test chamber placed against masonry with controlled temperature and humidity condition. Water is applied with increased air pressure and prime dampness observed.	Modified laboratory test. Specialized expertise required.	Provides useful information in quality of material coatings, leakage and workmanship. ASTM E514 can be adapted to in situ tests.	Expensive. Opposite side of wall may not be accessible to observe dampness.	2, 6, 8, 11, 13, 16, 17, 23, 25, 28, 30, 35

References for Table 3.7.10

1. ASTM C597. (1991). "Standard Test for Pulse Velocity Through Concrete," 1997 Annual Book of ASTM Standards, Vol. 4.02. American Society for Testing and Materials, West Conshohocken, Pennsylvania. 287–289.
2. ASTM E514. (1996). "Standard Test Method for Water Penetration and Leakage Through Masonry," 1997 Annual Book of ASTM Standards, Vol. 04.05. American Society for Testing and Materials, West Conshohocken, Pennsylvania, 1017–1020.
3. American Society of Metals. (1976). "Nondestructive Inspection and Quality Control, 8th Ed.," Vol. 11, American Society of Metals, Metal Park, Ohio.
4. Amrhein, J. E. (1983). "Reinforced Masonry Engineering Handbook," Masonry Institute of America, Los Angeles, California.
5. Atkinson, R. (1993). "Evaluating the Structural Condition of Existing Masonry," Concrete Repair Bulletin, March/April, 14–16.
6. Brown, M. T. (1990). "A Critical Review of Field Adapting ASTM E514 Water Permeability Test Method," Masonry: Components to Assemblages, *ASTM STP 1063*, edited by J. H. Matthys, American Society for Testing and Materials, West Conshohocken, Pennsylvania, 299–308.
7. Botsco, R. J. (1977). "Specialized NDT Methods—Lesson 12," Fundamentals of Nondestructive Testing, Metals Engineering Institute.
8. Farahmandpour, K., and Dubovoy, V. S. (1990). "Accelerated Field Test Method for Water Penetration of Masonry," *Proc., Non-Destructive Evaluation of Civil Structures and Materials*, edited by B. A. Suprenant, S. Sture, J. L. Noland, and M. P. Schuller, University of Colorado, Boulder, Colorado, October, 113–128.
9. Fattal, S. G., and Cattaneo, L. E. (1977). "Evaluation of Structural Properties of Masonry in Existing Buildings," BSS62, National Bureau of Standards, Washington, D.C.
10. Forrester, J. A. (1969). "Gamma Radiography of Concrete," *Proc., Symposium on Nondestructive Testing of Concrete and Timber*, Session No. 2, Institute of Civil Engineers, London, United Kingdom, 9–13.
11. Grimm, C. T. (1982a). "Water Permanence of Masonry Walls: A Review of the Literature," Masonry: Materials, Properties, and Performance, *ASTM STP 778*, edited by J. G. Borchelt, American Society for Testing and Materials, West Conshohocken, Pennsylvania, 178–199.
12. Grimm, C. T. (1982b). "Masonry Failure Investigations," Masonry: Materials, Properties, and Performance, *ASTM STP 778*, edited by J. G. Borchelt, American Society for Testing and Materials, West Conshohocken, Pennsylvania, 245–260.
13. Grimm, C. T. (1982c). "A Driving Rain Index for Masonry Walls," Masonry: Materials, Properties, and Performance, *ASTM STP 778*, edited by J. G. Borchelt, American Society for Testing and Materials, West Conshohocken, Pennsylvania, 245–260.

14. Grimmer, A. E. (1984). "A Glossary of Historic Masonry Deterioration Problems and Preservation Treatments," U.S. Department of Interior, National Park Service, Washington, D.C.
15. Hamid, A. A., and Drysdale, R. G. (1982). "Effect of Strain Gradient on Tensile Strength of Concrete Blocks," *Masonry: Materials, Properties, and Performance, ASTM STP 778*, edited by J. G. Borchelt, American Society for Testing and Materials, West Conshohocken, Pennsylvania, 57–65.
16. Krogstad, N. V. (1990). "Masonry Wall Drainage Test—A Proposed Method for Field Evaluation of Masonry Cavity Walls for Resistance to Water Leakage," *Masonry: Components to Assemblages, ASTM STP 1063*, edited by J. H. Matthys, American Society for Testing and Materials, West Conshohocken, Pennsylvania, 394–402.
17. Lerchen, F. H., Pielert, J. H., and Faison, T. K. (1980b). "Selected Methods for Condition Assessment of Structural, HVAC, Plumbing, and Electrical Systems in Existing Buildings," Section 2.4, Masonry, NBSIR 80–2171, National Bureau of Standards, Washington, D.C., 71–78.
18. Leslie, J. R., and Cheesman, W. J. (1949). "An Ultrasonic Method of Studying Deterioration and Cracking in Concrete Structures, Journal American Concrete Institute, 21(1), Vol. 46, Farmington Hills, Michigan, 17–36.
19. Livingston, R. A., Evans, L. G., Taylor, T. H., and Tomoka, J. I. (1986). "Diagnosis of Building Condition by Neutron Gamma Ray Technique," *Building Performance: Function, Preservation and Rehabilitation, ASTM STP 901*, ASTM, West Conshohocken, Pennsylvania.
20. Lorelace, J. F. (1977). "Ultrasonic Testing Equipment, Lesson 6," Fundamentals of Nondestructive Testing, Metals Engineering Institute.
21. Malhotra, V. M. (1976). "Testing Hardened Concrete: Nondestructive Methods," Monograph No. 9, American Concrete Institute, Farmington Hills, Michigan.
22. Moberg, A. J. (1977). "Ultrasonic Testing Applications, Lesson 7," Fundamentals of Nondestructive Testing, Metals Engineering Institute.
23. Monk, C. B., Jr. (1982). "Adaptations and Additions to ASTM Test Method E514 (Water Permeance of Masonry) for Field Conditions," Masonry: Materials, Properties, and Performance, *ASTM STP 778*, edited by J. G. Borchelt, American Society for Testing and Materials, West Conshohocken, Pennsylvania, 237–244.
24. Noland, J. L., Atkinson, R. H., and Schuller, M. P. (1990). "A Review of the Flatjack Method for Non-destructive Evaluation," *Proc., Non-Destructive Evaluation of Civil Structures and Materials*, edited by B. A. Suprenant, S. Sture, J. L. Noland, and M. P. Schuller, University of Colorado, Boulder, Colorado, October, 129–140.
25. Noland, J. L., Kingsley, G. R., and Atkinson, R. H. (1990). "Nondestructive Evaluation of Masonry: An Update," Masonry: Components to Assemblages, *ASTM STP 1063*, edited by J. H. Matthys, American Society for Testing and Materials, West Conshohocken, Pennsylvania, 248–262.
26. Parise, C. J. (1982). "Pulse Velocity Testing of Concrete," *Proc., American Society for Testing and Materials*, Vol. 53, West Conshohocken, Pennsylvania, 1033.
27. Parker, W. E. (1953). "Pulse Velocity Testing of Concrete," *Proc., ASTM*, Vol. 53, American Society for Testing and Materials, West Conshohocken, Pennsylvania, 1033.
28. Ribar, J. W. (1982). "Water Permeance of Masonry: A Laboratory Study," Masonry: Materials, Properties, and Performance, *ASTM STP 778*, edited by J. G. Borchelt, American Society for Testing and Materials, West Conshohocken, Pennsylvania, 200–220.
29. Rossi, P. P. (1990). "Non-Destructive Evaluation of the Mechanical Characteristics of Masonry Structures," *Proc., Non-Destructive Evaluation of Civil Structures and Materials*, edited by B. A. Suprenant, S. Sture, J. L. Noland, and M. P. Schuller, University of Colorado, Boulder, Colorado, October, 17–41.
30. Smith, B. M. (Updated). "Moisture Problems in Historic Masonry Walls," U.S. Department of Interior, National Park Service, Washington, D.C.
31. Smith, A. L. (1977). "Ultrasonic Testing Fundamentals, Lesson 5," Fundamentals of Nondestructive Testing, Metals Engineering Institute.
32. Snell, L. M. (1978). "Nondestructive Testing Techniques to Evaluate Existing Masonry Construction."
33. Stockbridge, J. G. (1983). "Techniques for Evaluating the Conditions of Masonry Buildings," *International Congress on Materials Science and Restoration*, Stuttgart, Germany.
34. Structural Clay Products Institute (now Brick Institute of America). (1969). "Causes and Control of Efflorescence on Brickwork," Research Report No. 15, Brick Institute of America, Reston, Virginia.
35. Suprenant, B. A. and Schuller, M. P. (1994). "Nondestructive Evaluation and Testing of Masonry Structures," The Aberdeen Group, Addison, Illinois.
36. U.S. Army. (1970). "Concrete and Masonry," Technical Manual No. 5-742, Department of the Army, Washington, D.C., June 22.
37. U.S. Department of Housing and Urban Development. (1982a). "Rehabilitation Guidelines, 9, Guideline for Structural Assessment, Chapter 3, Masonry Structures," Washington, D.C., 23–34.
38. Whitehurst, E. A. (1966). "Evaluation of Concrete Properties from Sonic Tests," American Concrete Institute, Monograph No. 2, Farmington Hills, Michigan.
39. Wilson, F. (1981). "Building Materials Evaluation Handbook," Van Nostrand Reinhold Company, New York, New York.

TABLE 3.7.11. Test Methods for Determining Mechanical and Physical Properties of Wood

Property	Purpose of Test	Test Method	User Expertise	Advantages	Limitations	References
Decay resistance	Evaluation of the natural decay resistance of wood.	Wood samples are exposed in decay chambers to pure cultures of decay fungi. Mass of sample is determined before and after exposure. Loss of mass is measure of decay.	Standard laboratory procedure. Moderate expertise required.	Useful in determining relative values of decay resistance.	Method is qualitative rather than quantitative. Samples from structure are required.	6
Elastic properties	Evaluation of module of elasticity, module of rigidity and Poisson's ratio.	Standard laboratory tests on clear and straight grained small pieces of wood.	Standard laboratory procedure. Moderate expertise required.	Useful in determining orthothropic properties for grading and determination of allowable properties.	Samples required from structure. Validity of results limited by number of specimens tested.	1, 2, 10, 14
Fasteners withdrawal resistance	Evaluation of resistance to withdrawal of nails, staples and screws.	Prisms of wood with fasteners driven into the specimen are tested for fastener withdrawal in a testing machine.	Standard laboratory procedure. Moderate expertise required.	Measures ability of fastener to hold adjoining members together.	Samples required from structure. Validity of results limited by number of specimens tested.	4
Fasteners mechanical properties	Determination of hardness, proof load, and tension of threaded fasteners.	Standard laboratory tests on fasteners.	Moderate expertise required.	Provides useful information needed for evaluation of connection.	Samples of fasteners must be removed from structure.	Table 3.7.6, 14
Moisture content	Determination of moisture content as percentage of oven-dry mass.	Mass of sample is determined, oven dried and the mass again determined.	Moderate expertise required.	Moisture content is one of most significant variables of wood affecting many properties significantly.	Samples must be taken from structure. Moisture content will vary with atmospheric condition and sample locations.	1, 5, 13
Preservatives	Determination of quality and effectiveness of wood preservatives.	Mass of sample is determined, preservative extracted and mass again determined. Test associated with moisture content test in which preservative content and moisture content are separated.	Moderate expertise required.	Preservative content is a measure of penetration and therefore service life of structure.	Samples must be taken from structure.	5, 7, 9

Specific gravity	Provides measure of amount of wood substance present in sample providing information on mass, workability and strength characteristics.	Value can be determined from moisture content tests or by water immersion or flotation methods.	Moderate expertise required.	Used to determined dead load of wood and for determining allowable properties.	Specific gravity is an indefinite quantity except for specific conditions under which it is determined. Samples required from structure.	8
Strength tests	Determination of orthotropic mechanical properties of specimens.	Test specimens are subjected to standard evaluation in testing machine using standard procedures.	Fairly high level of expertise required. Standard laboratory tests.	Used to determine structural grade and allowable properties of wood.	Samples must be taken from structure. Information limited by number of samples.	1, 2, 10
Structural grade and allowable properties	Grade and allowable properties needed to assess strength and life of structure.	Samples are classified by visual evaluation and from tests of mechanical properties.	Requires high level of expertise.	Provides useful information for structural assessment.	For good results, statistical approach must be used requiring a large number of samples from structure.	3, 11, 12, 14, 15

References for Table 3.7.11

1. ASTM D143. (1994). "Standard Methods of Testing Small Clear Specimens of Timber," 1997 Annual Book of ASTM Standards, Vol. 04.10. American Society for Testing and Materials, West Conshohocken, Pennsylvania, 23–53.
2. ASTM D198. (1994). "Standard Methods of Static Tests of Timbers in Structural Sizes," 1997 Annual Book of ASTM Standards, Vol. 04.10. American Society for Testing and Materials, West Conshohocken, Pennsylvania, 57–75.
3. ASTM D245. (1993). "Standard Practice for Establishing Structural Grades and Related Allowable Properties for Visually Graded Lumber," 1997 Annual Book of ASTM Standards, Vol. 04.10, American Society for Testing and Materials, West Conshohocken, Pennsylvania, 76–93.
4. ASTM D1761. (1995). "Standard Test Methods for Mechanical Fasteners in Wood," 1997 Annual Book of ASTM Standards, Vol. 4.10, American Society for Testing and Materials, West Conshohocken, Pennsylvania, 269–280.
5. ASTM D1860. (1995). "Standard Test Method for Moisture and Creosote-Type Preservative in Wood," 1997 Annual Book of ASTM Standards, Vol. 4.10, American Society for Testing and Materials, West Conshohocken, 285–287.
6. ASTM D2017. (1994). "Standard Method of Accelerated Laboratory Test of Natural Decay Resistance of Woods," 1997 Annual Book of ASTM Standards, Vol. 4.10, American Society for Testing and Materials, West Conshohocken, Pennsylvania, 313–317.
7. ASTM D2278. (1994). "Standard Test Method for Field Evaluation of Wood Preservatives in Round Post-Size Specimens," 1997 Annual Book of ASTM Standards, Vol. 4.10, American Society for Testing and Materials, West Conshohocken, Pennsylvania, 331–337.
8. ASTM D2395. (1993). "Standard Test Methods for Specific Gravity of Wood and Wood-Base Materials," 1997 Annual Book of ASTM Standards, Vol. 4.10, American Society for Testing and Materials, Philadelphia, West Conshohocken, 348–355.
9. ASTM D2481. (1994). "Standard Test Method for Accelerated Evaluation of Wood Preservatives for Marine Services by Means of Small Size Specimens," 1997 Annual Book of ASTM Standards, Vol. 4.10, American Society for Testing and Materials, West Conshohocken, Pennsylvania, 356–359.
10. ASTM D2555. (1996). "Standard Test Methods for Establishing Clear Wood Strength Values," 1997 Annual Book of ASTM Standards, Vol. 4.10, American Society for Testing and Materials, West Conshohocken, Pennsylvania, 360–375.
11. ASTM D2915. (1994). "Standard Practice for Evaluating Allowable Properties for Grades of Structural Lumber," 1997 Annual Book of ASTM Standards, Vol. 4.10, American Society for Testing and Materials, West Conshohocken, Pennsylvania, 398–409.
12. ASTM D3737. (1996). "Standard Practice for Establishing Stresses for Structural Glued Laminated Timber (Glulam)," 1997 Annual Book of ASTM Standards, Vol. 4.10, American Society for Testing and Materials, West Conshohocken, Pennsylvania, 454–471.
13. ASTM D4442. (1997). "Standard Test Methods for Direct Moisture Content Measurement of Wood and Wood-Base Materials," 1997 Annual Book of ASTM Standards, Vol. 4.10, American Society for Testing and Materials, West Conshohocken, Pennsylvania, 487–491.
14. ASTM F606. (1995b). "Standard Test Method for Determining the Mechanical Properties of Externally and Internally Threaded Fasteners, Washers, and Rivets," 1997 Annual Book of ASTM Standards, Vol. 15.08, American Society for Testing and Materials, West Conshohocken, Pennsylvania, 188–199.
15. U.S. Department of Agriculture. (1987). "Wood Handbook: Wood as an Engineering Material," Agriculture Handbook 72, Forest Products Laboratory, Forest Service, U.S. Department of Agriculture, Washington, D.C.

TABLE 3.7.12. NDT Methods for Determining Physical Conditions of Wood

Test Method	Purpose of Test	Principle of Operation	User Expertise	Advantages	Limitations	References
Coring	Coring can be used to determine the extent of decay and the thickness of char from fire damage. Cored samples can be sent to the laboratory to determine density, moisture content, type of decay and amount of preservative treatment.	An increment borer or cutting plug is used to extract a core. Core holes need to be treated and plugged to prevent decay in the future.	Moderate expertise is needed to determine core locations. A low level of expertise is needed to do the coring. Samples are taken to a laboratory for examination by a wood products pathologist.	A positive identification of the condition is made. Initial results are rapid.	Coring is time consuming. In confined areas it may be difficult or impossible to core. The extent of decay will require additional coring or drilling. Laboratory results are not immediately known.	6, 8, 13, 20, 32
Drilling	The reduced resistance to drilling indicates a possible pocket of decay. An examination of the drill chips and fines will confirm if decay has been encountered.	A member is drilled with an electric drill or a brace and bit. The depth when any change in resistance is encountered is recorded. The drill chips and fines are collected. Drill holes should be treated with a preservative.	A low level of expertise is required. Chips and fines can be taken to a laboratory for examination by a wood products pathologist.	Initial results are rapid. Procedure is easier than coring. Tool costs are inexpensive.	Extent of rot pocket may require numerous drill holes. Laboratory results are not immediately known.	20, 32
Moisture meter	Determination of moisture content of wood. The susceptibility of wood to decay is based in part on its moisture content. The strength of wood is also influenced by moisture. Moisture and its effects determine fiber saturation point and degree of distortion, twisting, warping, checking and splitting.	There are two types of moisture meters, dielectric and resistance. For resistance type meters, probes are driven into the wood and moisture content readings are recorded at varying depths. Dielectric meters do not require the driving of probes into the wood; the meter is simply held against the wood surface and an average moisture content at a predetermined depth is measured.	A low level of expertise is needed to take the readings.	Moisture content readings will indicate areas of decay or areas susceptible to decay. High readings indicate paths of moisture from leaking pipes, poor covering or poor flashing. The meter is small and portable.	Wood treated with creosote and pentachlorophenol can be tested with moisture meters with negligible error. Oxide formulations of CCA types B and C will cause slight errors but will still provide useful results. Meters should not be used with other types of chemical treatments. No meter is accurate above the fiber saturation point. (Approxi-	1, 2, 19, 20, 31, 35

Method	Description	Expertise/Advantages	Advantages	Limitations	Refs
			mately 26–28% of moisture content). Species and temperature influence the readings. Correction charts are supplied with moisture meters. A moisture meter relative in cost to a good camera must be purchased. Spare probe points and batteries should be stored with the meter.		
Probing	Probing with an awl or pick will locate surface areas that are soft and decayed. Sound wood that is pried will break into long splinters; decayed wood is brittle and will break into short splinters.	Low level of expertise required.	Results are rapid. Test can be done as part of a visual inspection. Costs are low.	Only the surface can be tested. Thick members will have to be tested by other methods to determine the condition of the interior.	6, 20, 26
Radiograph	Radiographs (X-rays) can be used to determine condition of highly inaccessible wood members; grain direction, irregularities, decay, splits, and insect damage.	X-rays are taken.	A high level of expertise is required.	The condition through the piece can be obtained with a nondestructive test. This is desirable in historic structures.	Results are not known until after the film is developed. The tests are costly. Access to both sides of the member is required.	14, 26, 45
Sounding	A poor hammer rebound or a dull or hollow sound indicates possible decay. Loose connections give off a hollow sound and may vibrate when struck with a hammer.	Surface is struck rapidly with changes in sound and rebound noted.	A high level of expertise is needed to interpret the sounds and rebound. Equipment is readily available and costs are low.	The test is rapid. Results are immediately known. Test can be performed during a visual inspection.	Coring or drilling will be required to determine the depth of decay. Internal decay in a thick member cannot be located.	1, 20

GUIDELINE FOR STRUCTURAL CONDITION ASSESSMENT OF EXISTING BUILDINGS

TABLE 3.7.12. NDT Methods for Determining Physical Conditions of Wood (*Continued*)

Test Method	Purpose of Test	Principle of Operation	User Expertise	Advantages	Limitations	References
Stress wave propagation	Sonic and ultrasonic techniques exist to detect and spatially quantify areas of advanced decay, insect and marine borer damage; to determine elastic properties and to estimate strength properties. Sonic and ultrasonic velocity and spectral content (frequency spectrum representation of stress wave time series) are influenced by density, moisture content, elastic properties, growth characteristics, processing variables and defects (including deterioration).	Dynamic elastic properties are directly calculated from velocity and wood density. Estimates of static elastic and strength properties can be obtained from measured dynamic elastic properties. Spectral content can be used to estimate strength properties. Estimates of static elastic and strength properties using either velocity or spectral content depend on empirical relationships. Sudden reductions in velocities through the thickness of a member or shifts in spectral content are indicators of void, advanced decay, insect damage, etc.	Moderate to high expertise is required depending on the equipment and technique employed. Recommended test and data analysis procedures must be carefully followed to obtain reliable results. Results from other types of tests such as visual inspection and sound-and-bore are typically evaluated along with stress wave results, requiring inspector judgement.	Elastic properties can be determined with a high level of accuracy. Excellent methods for initial screening of members for presence of advanced decay and voids. Some techniques offer immediate, on-site estimates of mechanical properties. Most equipment is portable and test times range from quick to moderately long depending on methods used and desired accuracy. The use of stress wave techniques in combination with visual and/or sound-and-bore techniques offers an excellent compromise between test complexity and accuracy of results.	Accurate assessment of elastic properties depends on knowing wood density. Velocity and spectral content are influenced by wood moisture content, temperature, chemical treatment, and orientation of grain and growth rings to the direction of stress wave propagations, which can lead to errors in interpreting test data. Only modest accuracy can be obtained in spatial quantification of decay and voids and in estimation of strength properties.	3, 4, 5, 6, 9, 10, 11, 12, 15, 16, 17, 18, 22, 23, 24, 26, 27, 29, 33, 34, 36, 37, 38, 39, 42, 45, 47
Visual and visual stress grading	A visual inspection is used to determine the appearance and visible condition of the member. If a grade stamp is not visible, then the physical dimensions of the member, size and location of growth characteristics are used to determine the grade. The cause of failure may be evident by the appearance of the failed members.	Visually examine all exposed surfaces and record all dimensions and conditions found: for example, knots, checks, splits, slope of grain, stains and connection details. Record all details of fire damaged and failed members. Consider video taping, tape recording and/or photographing during a visual inspection.	The inspector must have knowledge of wood as a structural material and how it responds to acceptable and adverse loadings and to its environment.	Results are usually rapid. A general inspection (walk through) can be done quickly and a trained eye will know where to look for visible damage. A survey will take longer. Cost is principally for manpower, not equipment.	Only visible surfaces can be inspected. Finishes may have to be removed to view hidden surfaces. Load tests may be required for systems not easily exposed. Trained personnel are required.	1, 6, 7, 13, 20, 21, 26, 28, 30, 38, 40, 41, 43, 44, 45, 46, 48

84

References for Table 3.7.12

1. ASCE Technical Committee on Wood. (1982). "Evaluation, Maintenance and Upgrading of Wood Structures, A Guide and Commentary," prepared by the Subcommittee on Evaluation, Maintenance and Upgrading of Timber Structures under the Technical Committee on Wood. ASCE, Reston, Virginia.
2. ASTM D4444. (1992). "Standard Test Methods for Use and Calibration of Hand-Held Moisture Meters," 1997 Annual Book of ASTM Standards, Vol. 4.10, American Society for Testing and Materials, West Conshohocken, Pennsylvania, 492–497.
3. Aggour, M. S., and Porbahai, M. A. (1989). "Underwater Strength of Timber Piles," *Proc., Int. Conf. on Wood Poles and Piles*, Colorado State University and Engineering Data Management, Inc., Fort Collins, Colorado, October 25–27, H1–H9.
4. Anthony, R. W., and Bodig, J. (1989). "Nondestructive Evaluation of Timber Structures for Reliable Performance," *Proc., Second Pacific Timber Engineering Conference*, University of Auckland, Auckland, New Zealand, August 28–31, Vol. 2, 197–200.
5. Anthony, R. W., and Bodig, J. (1990). "Applications of Stress Wave-based Non-Destructive Evaluation to Wood," *Proc., Non-Destructive Evaluation of Civil Structures and Materials*, edited by B. A. Suprenant, S. Sture, J. L. Noland, and M. P. Schuller, University of Colorado, Boulder, Colorado, October, 257–275.
6. Anthony, R. W., and Phillips, G. E. (1989). "Nondestructive Strength Assessment of In Situ Timber Piles," *Proc., International Conference on Wood Poles and Piles*, Colorado State University and Engineering Data Management, Inc., Fort Collins, Colorado, October 25–27, H10–H23.
7. Bendtsen, A., and Galligan, W. L. (1978). "Deriving Allowable Properties of Lumber (A Practical Guide for Interpretation of ASTM Standards)," FPL 20, Forest Products Laboratory, Forest Service, U.S. Department of Agriculture, Madison, Wisconsin.
8. Bodig, J., and Phillips, G. E. (1990). "Quantification of Biodegradation in Wood of Inaccessible Locations, Phase 1," Final Report to the National Science Foundation, Small Business Innovation Research Program, Grant No. ISI-8960992, Fort Collins, Colorado.
9. Brooks, R. T., and Burk, H. G. (1992). "Determining In Situ Timber Pile Length Using Stress Waves," Final Report to the Timber Bridge Initiative Special Projects Program by Engineering Data Management, Inc., Fort Collins, Colorado.
10. Brooks, R. T., and Phillips, G. E. (1991). "Nondestructive Strength Assessment of Wood Cooling Tower Columns," *Proc., Eighth International Nondestructive Testing of Wood Symposium*, Washington State University, Vancouver, Washington, September 23–25, 219–233.
11. Browne, C. M., and Kuchar, W. E. (1985). "Determination of Material Properties for Structural Evaluation of Trestle," *Proc., Fifth Nondestructive Testing of Wood Symposium*, Pullman, Washington, September 9–11, Conferences and Institutes, Washington State University, Pullman, Washington, 361–384.
12. Chui, Y. H. (1989). "Vibration Testing of Wood and Wooden Structure—Practical Difficulties and Possible Sources of Error," *Proc., Seventh International Nondestructive Testing of Wood Symposium*, Washington State University, Pullman, Washington, September 27–29, 173–188.
13. Core, H. A., and Cote, W. A. (1979). "Wood Structure and Identification," 2nd edition, Syracuse University Press, Syracuse, New York.
14. Hart, D. M. (1977). "X-Ray Inspection of Historic Structures: An Aid to Dating and Structural Analysis," *Technology and Conservation*, Vol. 2, No. 2, 10–13, 27.
15. Hearmon, R. F. S. (1966). "Theory of the Vibration Testing of Wood," Forest Products Journal, Vol. 16, No. 8, 29–40.
16. Hoyle, R. J., and Pellevin, R. F. (1978). "Stress Wave Inspection of a Wood Structure." *Fourth Symposium on Nondestructive Testing of Wood*, Washington State University, Pullman, Washington, August 28–30, 33–45.
17. Hoyle, R. J., and Rutherford, P. S. (1987). "Stress Wave Inspection of Bridge Timbers and Decking," Washington State Transportation Center, WSDOT Technical Monitor, Dean Moberg Bridge Condition Survey Office, Washington State Department of Transportation, Olympia, Washington.
18. James, W. L. (1961). "Internal Friction and Speed of Sound in Douglas-fir as Affected by Temperature and Moisture Content," Forest Products Laboratory, Forest Service, U.S. Department of Agriculture, Madison, Wisconsin.
19. James, W. L. (1988). "Electric Moisture Meters for Wood," General Technical Report FPL-GTR-6, Forest Products Laboratory, Forest Service, U.S. Department of Agriculture, Madison, Wisconsin.
20. Janey, J. R. (1986). "Guide to Investigation of Structural Failure," Report prepared for ASCE Research Council on Performance of Structures, ASCE, Reston, Virginia.
21. Jedrzejewski, M. J. (1986). "A Method for Determining the Allowable Strength of In-Place Wood Structural Members," Building Performance: Function, Preservation, and Rehabilitation, *ASTM STP 901*, edited by G. Davis, American Society for Testing and Materials, West Conshohocken, Pennsylvania, 136–151.
22. Kaiserlik, J. (1978). "Nondestructive Testing Methods to Predict Effect of Degradation on Wood: A Critical Assessment," General Technical Report FPL 19, Forest Products Laboratory, Forest Service, U.S. Department of Agriculture, Madison, Wisconsin.
23. Lanius, R. M., Jr., Tichy, R., and Bullet, W. M. (1981). "Strength of Old Wood Joists," Journal of the Structural Division, ASCE, Vol. 107, No. 12, Paper No. 16767, December, 2349–2363.
24. Lee, I. D. G. (1970). "Testing for Safety in Timber Structures," *Symposium on Non-Destructive Testing of Concrete and Timber*, The Institute of Civil Engineers, London, U.K.
25. Lemaster, R. L., and Dornfeld, D. A. (1987). "Preliminary Investigation of the Feasibility of Using Acousto-Ultrasonics to Measure Defects in Lumber," Journal of Acoustic Emission, Vol. 6, No. 3, 157–165.

References for Table 3.7.12 (Continued)

26. Lerchen, F. H., Pielert, J. H., and Faison, T. K. (1980a). "Selected Methods for Condition Assessment of Structural, HVAC, Plumbing, and Electrical Systems in Existing Buildings," Section 2.2. Wood (Structural Timber), NABSIR 80-2171, National Bureau of Standards, Washington, D.C., 39–55.
27. McDonald, K. A. (1978). "Lumber Defect Detection by Ultrasonics," Research Paper FPL 311, Forest Products Laboratory, Forest Service, U.S. Department of Agriculture, Madison, Wisconsin.
28. Naval Facilities Engineering Command (NFEC). (1985). "Inspection of Wood Beams and Trusses," NAVFAC MD-111.1, Arlington, Virginia.
29. Neal, D. W. (1985). "Establishment of Elastic Properties for In-Place Timber Structures," *Fifth Nondestructive Testing of Wood Symposium*, Pullman, Washington, September 9–11, Conferences and Institutes, Washington State University, Pullman, Washington, 353–359.
30. Northeast Lumber Manufacturers Association. (1990). "Standard Grading Rules for Northeast Lumber," P.O. Box 87A, Cumberland Center, Maine 04021.
31. Panshin, A. J., and deZeeuw, C. (1970). "Textbook of Wood Technology, Volume 1," McGraw-Hill Book Company, New York, New York.
32. Patton-Mallory, M., and DeGroot, R. C. (1989). "Detecting Brown-Rot Decay in Southern Yellow Pine by Acousto-Ultrasonics," *Proc., Seventh International Nondestructive Testing of Wood Symposium*, Washington State University, Pullman, Washington, September 27–29, 28–44.
33. Pellerin, R. F. (1989). "Inspection of Wood Structures for Decay Using Stress Waves," *Proc., Second Pacific Timber Engineering Conference*, University of Auckland, Auckland, New Zealand, August 28–31, Vol. 2, 191–195.
34. Pellerin, R. F., DeGroot, R. C., and Esenther, G. R. (1985). "Nondestructive Stress Wave Measurements of Decay and Termite Attack in Experimental Wood Units," *Proc., Fifth Nondestructive Testing of Wood Symposium*, Washington State University, Pullman, Washington, September 9–11, 319–352.
35. Richards, M. J. (1990). "Effect of CCA-C Wood Preservative on Moisture Content Readings by The Electronic Resistance—Type Moisture Meter," Forest Products Journal, Vol. 4, No. 2, 29–33.
36. Ross, R. J., and Pellerin, R. F. (1994). "Nondestructive Testing for Assessing Wood Members in Structures: A Review," General Technical Report FPL-GTR-70, Forest Products Laboratory, Forest Service, U.S. Department of Agriculture, Madison, Wisconsin.
37. Rutherford, P. S., Hoyle, R. J., Jr., DeGroot R. C., and Pellerin, R. F. (1987). "Dynamic vs. Static MOE in the Transverse Direction in Wood," *Proc., Sixth Nondestructive Testing of Wood Symposium*, Washington State University, Pullman, Washington, September 14–16, 67–80.
38. Sandoz, J. L. (1990). "Non-destructive Evaluation in Wood Structures: Tools and Reliability Concepts," *Proc., Non-Destructive Evaluation of Civil Structures and Materials*, edited by B. A. Suprenant, S. Sture, J. L. Noland, and M. P. Schuller, University of Colorado, Boulder, Colorado, October, 43–59.
39. Sandoz, J. L. (1991). "Nondestructive Evaluation of Building Timber by Ultrasound," *Eighth International Nondestructive Testing of Wood Symposium*. Washington State University, Vancouver, Washington, September 23–25. 131–142.
40. Soltis, L. A. (1986). "Details for Durable Timber Structures," *Effects of Deterioration on Safety and Reliability of Structures, Proc., Session sponsored by the Structural Division of the American Society of Civil Engineers in Conjunction with the ASCE Convention*, edited by S. M. Ma, Seattle, Washington, April 10, 12–21.
41. Southern Pine Inspection Bureau. (1990). "Grading Rules," 4709 Scenic Highway, Pensacola, Florida, 32504.
42. Stewart, A. H., Brunette, T. L., and Goodman, J. R., Jr. (1986). "Use of Nondestructive Testing in Rehabilitation of Wood Cooling Towers," Evaluation and Upgrading of Wood Structures, Case Studies, *Proc., Session at Structures Congress '86 sponsored by the Structural Division*, edited by V. K. A. Gopu, New Orleans, Louisiana, September 15–18, American Society of Civil Engineers, Reston, Virginia, 13–27.
43. U.S. Department of Commerce. (1986). "American Softwood Lumber Standard," Voluntary Product Standard PS20–70, Amended. National Bureau of Standards, Washington, D.C., 1–26.
44. U.S. Department of Commerce. (1973). "Structural Glued Laminated Timber," Voluntary Product Standard PS56–73, National Bureau of Standards, Washington, D.C., 1–10.
45. U.S. Department of Housing and Urban Development. (1982a). "Rehabilitation Guidelines, 9, Guideline for Structural Assessment, Chapter 2, Wood and Timber Structures," Washington, D.C., 16–22.
46. U.S. Department of Housing and Urban Development. (1984). "Rehabilitation Guidelines, 11, Guideline for Residential Building Systems Inspection, Chapter 4.5, Structural System, Wood Structural Components," Washington, D.C., 57–63.
47. Volney, N. J. (1991). "Timber Bridge Inspection Case Studies in Use of Stress Wave Velocity Equipment," *Proc., Eighth International Nondestructive Testing of Wood Symposium*, Washington State University, Vancouver, Washington, September 23–25, 235–246.
48. West Coast Lumber Inspection Bureau. (1991). "West Coast Lumber Grading Rules," P.O. Box 23145, Portland, Oregon 97223.

SEI/ASCE 11-99

TABLE 3.7.13. Test Methods for Load Testing on Buildings

Test Method	Purpose of Test	Principle of Operation	User Expertise	Advantages	Limitations	References
Analysis and interpretation of data	Limit states criteria, proof loads and structure reliability.	Mathematical analysis of load test data using statistical interpretation when applicable.	Highly qualified engineer required to perform analysis and interpretations.	With qualified analysis and interpretation, types and locations of defects can be determined as well as load capacities of members and systems.	Limited by amount and type of data recorded—whether strains deflections or dynamic responses—and hence on amount of instrumentation.	2, 3, 16, 18
Dynamic load tests	Mainly applicable to system buildings.	An existing force is applied to a member or structural system and the structure instrumented to measure dynamic response.	Highly qualified engineer required to design and supervise load tests.	Dynamic signature cannot only be helpful in identifying defects, but provides data base for future testing to identify changes in structure.	Dynamic loading of systems is expensive and requires specially made equipment to apply exciting force.	6, 8, 11, 12, 23, 26, 28, 30, 31
Monitoring dynamic response	Usually permanent systems to monitor seismic activity.	Instrument critical location and members. Use velocity transducers, accelerometers or seismometers.	Highly qualified engineer required.	Structure signature is obtained for seismic loads.	Not very applicable in identifying specific deterioration problems.	26, 30
Monitoring strain/stress	Strains are recorded under service loads (and sometimes for specific static loads).	Instrument members and measure strains at critical locations. Use of strain gauges.	Highly qualified engineer required.	Results over a long time period can be recorded automatically. Results can pinpoint over-stressed members.	Expensive to establish enough strain gauges to obtain meaningful results.	25, 27
Static load tests	In situ load to predict how members or system will perform. Can generally determine only load capacity based on bending.	Specific and identifiable loads corresponding to multiples of the service load are applied and deflections/strains recorded for the severe load applications.	Highly qualified engineer required.	Can determine actual proof load independent of analysis and interpretation. Most highly used method by load testing and therefore advanced development of procedures is available.	Static load testing is expensive.	1, 2, 4, 5, 7, 9, 10, 13, 14, 15, 17, 19, 20, 21, 22, 24, 28, 29, 32, 33

References for Table 3.7.13
1. ACI Committee 318. (1995). "Building Code Requirements for Reinforced Concrete and Commentary, Chapter 20: Strength Evaluation of Existing Structures," ACI Designations 318-95 and ACI 318R-95, American Concrete Institute, Farmington Hills, Michigan, 297–300.
2. ACI Committee 437. (1995). "Strength Evaluation of Existing Concrete Buildings," ACI Designation 437R-91, American Concrete Institute, Farmington Hills, Michigan, 24 pp.
3. Allen, D. E. (1991). "Limit States Criteria for Structural Evaluation of Existing Buildings," Canadian Journal of Civil Engineering, CSCE, Vol. 18, 995–1004.
4. BOCA. (1993). "The BOCA National Building Code, Article 11, Structural Loads, Sections 1102.3 and 1102.4 and Article 13, Materials and Tests," Building Officials & Code Administrators International, Inc., Country Club Hills, Illinois.
5. Bares, R., and Fitzsimons, N. (1975). "Load Tests of Building Structures," Journal of the Structural Division, ASCE, Vol. 101, No. 5, Proc. Paper 11322, May, 1111–1123.

References for Table 3.7.13 (Continued)

6. Blasi, C., Spinelli, P., and Vignoli, A. (1987). "Dynamic Behavior and Monitoring of Ancient Monuments," *IABSE Reports*, Vol. 56, International Association for Bridge and Structural Engineering, Zurich, Switzerland, September, 311–324.
7. Cangiano, M., and Croci, G. (1987). "Control and Detecting System Related to the Construction of the Underground Railway Line in Rome," *IABSE Colloquium, Bergamo, Monitoring Large Structures and Assessment of their Safety, IABSE Reports*, Vol. 56, International Association for Bridge and Structural Engineering, Zurich, Switzerland, September, 339–355.
8. Castoldi, A., Chiarugi, A., Giuseppetti, G., and Fanelli, M. (1987). "In Situ Dynamic Tests on Ancient Monuments," *IABSE Colloquium, Bergamo, Monitoring Large Structures and Assessment of their Safety, IABSE Reports*, Vol. 56, International Association for Bridge and Structural Engineering, Zurich, Switzerland, September, 131–144.
9. Castoldi, A., Chiarugi, A., Giuseppetti, G., and Petrini, G. (1987). "Static Monitoring System for the Brunelleschi Dome in Florence," *IABSE Colloquium, Bergamo, Monitoring Large Structures and Assessment of their Safety, IABSE Reports*, Vol. 56, International Association for Bridge and Structural Engineering, Zurich, Switzerland, September, 17–33.
10. Colville, J. (1980). "Instrumentation and Testing of a Full-Scale Masonry Structure," *Symposium of Full-Scale Load Testing of Structures, ASTM STP 702*, edited by W. R. Schriever, American Society for Testing and Materials, West Conshohocken, Pennsylvania, 114–124.
11. Corti, M., Marazzini, S., Martinelli, A., and De Agostini, A. (1987). "Cross-Spectrum Technique for High-Sensitivity Remote Vibration Analysis by Optical Interferometry," *IABSE Colloquium, Bergamo, Monitoring Large Structures and Assessment of their Safety, IABSE Reports*, Vol. 56, International Association for Bridge and Structural Engineering, Zurich, Switzerland, September, 281–291.
12. Croci, G. (1987). "The Role of Monitoring for the Knowledge of the Behaviour of Structures," *IABSE Colloquium, Bergamo, Monitoring Large Structures and Assessment of their Safety, IABSE Reports*, Vol. 56, International Association for Bridge and Structural Engineering, Zurich, Switzerland, September, 35–48.
13. Dixon, D. E., and Smith, J. R. (1980). "Skyline Plaza North (Building A-4)—A Case Study," *Symposium of Full-Scale Load Testing of Structures, ASTM STP 702*, edited by W. R. Schriever, American Society for Testing and Materials, West Conshohocken, Pennsylvania, 182–199.
14. Fitzsimons, N., and Longinov, A. (1975). "Guidance for Load Tests of Buildings," *Journal of the Structural Division*, ASCE, Vol. 101, No. 7, Proc. Paper 11431, July, 1367–1380.
15. Fritz, T. W. (1980). "Structural Testing of Mobile-Home Roof/Ceiling Assemblies," *Symposium of Full-Scale Load Testing of Structures, ASTM STP 702*, edited by W. R. Schriever, American Society for Testing and Materials, West Conshohocken, Pennsylvania, 135–147.
16. Fujino, Y., and Lind, N. C. (1977). "Proof-Load Factors and Reliability," *Journal of The Structural Division*, ASCE, Vol. 103, No. 4, Proc. Paper 12885, April, 853–870.
17. Guedelhoefer, O. C., and Janney, J. R. (1980). "Evaluation of Performance by Full-Scale Testing," *Symposium of Full-Scale Load Testing of Structures, ASTM STP 702*, edited by W. R. Schriever, American Society for Testing and Materials, West Conshohocken, Pennsylvania, 5–24.
18. Hall, W. B. (1988). "Reliability of Service—Proven Structures," *Journal of Structural Engineering*, ASCE, Vol. 114, No. 3, Paper No. 22275, March, 608–624.
19. Harris, H. G. (1980). "Use of Structural Models as an Alternative to Full Scale Testing," *Symposium on Full-Scale Load Testing of Structures, ASTM STP 702*, edited by W. R. Schriever, American Society for Testing and Materials, West Conshohocken, Pennsylvania, 25–44.
20. Ivanyi, M. (1976). "Discussion of 'Load Tests of Building Structures,'" by Richard Bares and Neal Fitzsimons, *Journal of the Structural Division*, ASCE, Vol. 102, No. 6, Proc. Paper 12166, June, 1260–1261.
21. Johnson, D. L. (1980). "A Method for the Full-Scale Testing of Roof Systems," *Symposium of Full-Scale Load Testing of Structures, ASTM STP 702*, edited by W. R. Schriever, American Society for Testing and Materials, West Conshohocken, Pennsylvania, 78–87.
22. Longinow, A., and Schriever, W. R. (1980). "Conclusions from July 1976 Engineering Foundation Conference in Rindge, N.H. on Full-Scale Structural Testing," *Symposium of Full-Scale Load Testing of Structures, ASTM STP 702*, edited by W. R. Schriever, American Society for Testing and Materials, West Conshohocken, Pennsylvania, 169–181.
23. Radogna, E. F., Falletti, E., Da Rin, E. M., Rossi, U., and Stefani, S. (1987). "Full-Scale Dynamic Testing of ENEL Power Plant Structures," *IABSE Colloquium, Bergamo, Monitoring Large Structures and Assessment of their Safety, IABSE Reports*, Vol. 56, International Association for Bridge and Structural Engineering, Zurich, Switzerland, September, 227–236.
24. Raths, C. H., and Guedelhoefer, O. C. (1980). "Correlation of Load Testing with Design," *Symposium of Full-Scale Load Testing of Structures, ASTM STP 702*, edited by W. R. Schriever, American Society for Testing and Materials, West Conshohocken, Pennsylvania, 91–113.
25. Russell, H. G. (1980). "Field Instrumentation of Concrete Structures," *Symposium of Full-Scale Load Testing of Structures, ASTM STP 702*, edited by W. R. Schriever, American Society for Testing and Materials, West Conshohocken, Pennsylvania, 63–77.
26. Soong, T. T. (1990). "Active Structural Control of Building Structures," *Proc., Bridges and Buildings of the 90's*, ASCE, Metropolitan Section, The Structures Group, Spring Seminar, B89-B98.
27. Sørum, G., and Dyken, T. (1987). "Vibrating-Wire Reinforcement Strain Gauges for Performance Monitoring of Large Concrete Structures," *IABSE Colloquium, Bergamo, Monitoring Large Structures and Assessment of their Safety, IABSE Reports*, Vol. 56, International Association for Bridge and Structural Engineering, Zurich, Switzerland, September, 83–93.
28. Standard Building Code. (1994). "Chapter 17, Structural Tests and Inspections," Southern Building Code Congress International, Birmingham, Alabama.
29. Tuomi, R. L. (1980). "Full-Scale Testing of Wood Structures," *Symposium of Full-Scale Load Testing of Structures, ASTM STP 702*, edited by W. R. Schriever, American Society for Testing and Materials, West Conshohocken, Pennsylvania, 45–62.
30. Uniform Building Code. (1991). "Appendix: Division II, Earthquake Recording Instrumentation," Uniform Building Code, International Conference of Building Officials, Whitter, California, May, 874–875.
31. Vestroni, F., and Capecchi, D. (1987). "Aspects of the Application of Structural Identification in Damage Evaluation," *IABSE Colloquium, Bergamo, Monitoring Large Structures and Assessment of their Safety, IABSE Reports*, Vol. 56, International Association for Bridge and Structural Engineering, Zurich, Switzerland, September, 83–93.
32. West, R. E., Osborn, A. E. N., and Rentschler, G. P. (1987). "Structural Monitoring of a Large Space Frame," *IABSE Colloquium, Bergamo, Monitoring Large Structures and Assessment of their Safety, IABSE Reports*, Vol. 56, International Association for Bridge and Structural Engineering, Zurich, Switzerland, September, 215–226.
33. Yancey C. W. C. (1980). "The Use of Full-Scale Testing for Establishing and Validating Structural Performance Requirements," *Symposium of Full-Scale Load Testing of Structures, ASTM STP 702*, edited by W. R. Schriever, American Society for Testing and Materials, West Conshohocken, Pennsylvania, 148–168.

3.8 SUPPLEMENTAL REFERENCES

1. ASTM E455. (1998). "Standard Method for Static Load Testing of Framed Floor or Roof Diaphragm Constructions for Buildings," 1998 Annual Book of ASTM Standards, *Vol. 04.11*, American Society for Testing and Materials, West Conshohocken, Pennsylvania, 52–56.
2. ASTM E529. (1994). "Standard Methods of Flexural Tests on Beams and Girders for Building Construction," 1998 Annual Book of ASTM Standards, *Vol. 04.11*, American Society for Testing and Materials, West Conshohocken, Pennsylvania, 71–73.
3. Alampalli, S., Fu, G., and Aziz, I. A. (1992). "Nondestructive Evaluation of Highway Bridges by Dynamic Monitoring," *Proc., Nondestructive Evaluation of Civil Structures and Materials*, edited by B. A. Suprenant, J. L. Noland, and M. P. Schuller, University of Colorado, Boulder, Colorado, May, 211–225.
4. Alexander, A. M. (1992). "Accuracy of Predicting In Situ Compressive Strength of Deteriorated Concrete Seawall by NDT Methods," *Proc., Nondestructive Evaluation of Civil Structures and Materials*, edited by B. A. Suprenant, J. L. Noland, and M. P. Schuller, University of Colorado, Boulder, Colorado, May, 68–82.
5. Alexander, A. M., and Thornton, H. J., Jr. (1988). "Developments in Ultrasonic Pitch-Catch and Pulse-Echo for Measurements in Concrete," Nondestructive Testing, *ACI Publication SP-112*, edited by H. S. Lew, American Concrete Institute, Farmington Hills, Michigan, 21–40.
6. Al-Qadi, I. L. (1992). "The Penetration of Electromagnetic Waves into Hot-Mix Asphalt," *Proc., Nondestructive Evaluation of Civil Structures and Materials*, edited by B. A. Suprenant, J. L. Noland, and M. P. Schuller, University of Colorado, Boulder, Colorado, May, 195–209.
7. Anthony, R. W. (1992). "Structural Timber Assessment Using Nondestructive Evaluation," *Proc., Nondestructive Evaluation of Civil Structures and Materials*, edited by B. A. Suprenant, J. L. Noland, and M. P. Schuller, University of Colorado, Boulder, Colorado, May, 357–373.
8. Berra, M., Binda, L., Anti, L., and Fatticcioni, A. (1992). "Utilization of Sonic Tests to Evaluate Damaged and Repaired Masonries," *Proc., Nondestructive Evaluation of Civil Structures and Materials*, edited by B. A. Suprenant, J. L. Noland, and M. P. Schuller, University of Colorado, Boulder, Colorado, May, 329–338.
9. Bruns, S. R., and Higgins, C. C. (1992). "Evaluation and Testing of a Closed Spandrel Reinforced Concrete Arch Bridge," *Proc., Nondestructive Evaluation of Civil Structures and Materials*, edited by B. A. Suprenant, J. L. Noland, and M. P. Schuller, University of Colorado, Boulder, Colorado, May, 395–410.
10. Buchanan, D. J. (1992). "Use of Hydrophones to Monitor Prestressed Concrete Pipe," *Proc., Nondestructive Evaluation of Civil Structures and Materials*, edited by B. A. Suprenant, J. L. Noland, and M. P. Schuller, University of Colorado, Boulder, Colorado, May, 427–435.
11. Bungey, J. H. (1984). "The Influence of Reinforcement on Ultrasonic Pulse Velocity Testing," In Situ/Nondestructive Testing of Concrete, *ACI Publication SP-82*, edited by V. M. Malhotra, American Concrete Institute, Farmington Hills, Michigan, 229–246.
12. Cantor, T. R. (1984). "Review of Penetrating Radar as Applied to Nondestructive Evaluation of Concrete," In Situ/Nondestructive Testing of Concrete, *ACI Publication SP-82*, edited by V. M. Malhotra, American Concrete Institute, Farmington Hills, Michigan, 581–601.
13. Carino, N. J. (1984). "Laboratory Study of Flaw Detection in Concrete by the Pulse-Echo Method," In Situ/Nondestructive Testing of Concrete, *ACI Publication SP-82*, edited by V. M. Malhotra, American Concrete Institute, Farmington Hills, Michigan, 557–579.
14. Carlsson, M., Eeg, I. R., Jahren, P. (1984). "Field Experience in the Use of the 'Break-Off Tester,'" In Situ/Nondestructive Testing of Concrete, *ACI Publication SP-82*, edited by V. M. Malhotra, American Concrete Institute, Farmington Hills, Michigan, 277–292.
15. Clark, W. G., Jr., and Shannon, R. E. (1992). "Magnetic Tagging for the Improved Characterization of Construction Materials," *Proc., Nondestructive Evaluation of Civil Structures and Materials*, edited by B. A. Suprenant, J. L. Noland, and M. P. Schuller, University of Colorado, Boulder, Colorado, May, 113–127.
16. Cohen, J. S., and Osborn, A. E. N. (1992). "Strain Gage Monitoring for the Queens Midtown Viaduct," *Proc., Nondestructive Evaluation of Civil Structures and Materials*, edited by B. A. Suprenant, J. L. Noland, and M. P. Schuller, Uni-

versity of Colorado, Boulder, Colorado, May, 453–470.
17. Dahl-Jorgensen, E., and Johansen R. (1984). "General and Specialized Use of the Break-Off Concrete Strength Testing Method," In Situ/Nondestructive Testing of Concrete, *ACI Publication SP-82*, edited by V. M. Malhotra, American Concrete Institute, Farmington Hills, Michigan, 293–308.
18. Davis, A. G., and Paquet, J. (1992). "Monitoring Bridge Performance Using NDE Techniques," *Proc., Nondestructive Evaluation of Civil Structures and Materials*, edited by B. A. Suprenant, J. L. Noland, and M. P. Schuller, University of Colorado, Boulder, Colorado, May, 227–240.
19. Dilly, R. L., and Vogt, W. L. (1988). "Pullout Test, Maturity, and PC Spreadsheet Software," Nondestructive Testing, *ACI Publication SP-112*, edited by H. S. Lew, American Concrete Institute, Farmington Hills, Michigan, 193–218.
20. Erlin, B. (1990). "The Magic of Investigative Petrography: The Practical Basis for Resolving Concrete Problems," Petrography Applied to Concrete and Concrete Aggregates, *ASTM STP 1061*, edited by B. Erlin and D. Stark, American Society for Testing and Materials, West Conshohocken, Pennsylvania, 171–181.
21. Figg, J. W., Sheehan, A., and Tomsett, H. N. (1992). "Assessment of Composite Concrete Floors Using a Combination of Sonic and Ultrasonic Non-destructive Test Methods," *Proc., Nondestructive Evaluation of Civil Structures and Materials*, edited by B. A. Suprenant, J. L. Noland, and M. P. Schuller, University of Colorado, Boulder, Colorado, May, 145–160.
22. Forde, M. C. (1992). "Non-Destructive Evaluation of Masonry Bridges," *Proc., Nondestructive Evaluation of Civil Structures and Materials*, edited by B. A. Suprenant, J. L. Noland, and M. P. Schuller, University of Colorado, Boulder, Colorado, May, 17–42.
23. Gibson, R. F. (1985). "Frequency Domain Testing of Materials," *Proc., Fifth Nondestructive Testing of Wood Symposium*, Pullman, Washington, September 9–11, Conferences and Institutes, Washington State University, Pullman, Washington, 385–406.
24. Ghorbanpoor, A., McGogney, C. H., and Virmani, Y. P. (1992). "Recent Developments in NDE of Concrete Bridge Structures," *Proc., Nondestructive Evaluation of Civil Structures and Materials*, edited by B. A. Suprenant, J. L. No-

land, and M. P. Schuller, University of Colorado, Boulder, Colorado, May, 247–261.
25. Greimann, L., and Stecker, J. (1990). "Maintenance and Repair of Steel Sheet Pile Structures," *Technical Report REMR-OM-9*, U.S. Army Construction Engineering Research Laboratory, Champaign, Illinois, December, 140 pp.
26. Harrell, T. R. (1988). "Nondestructive Testing Use in Evaluating the Form Removal Time for a Large Diameter Tunnel Concrete Lining," Nondestructive Testing, *ACI Publication SP-112*, edited by H. S. Lew, American Concrete Institute, Farmington Hills, Michigan, 153–164.
27. Hearn, G., and Ghia, R. (1992). "Response-Based Structural Condition Monitoring," *Proc., Nondestructive Evaluation of Civil Structures and Materials*, edited by B. A. Suprenant, J. L. Noland, and M. P. Schuller, University of Colorado, Boulder, Colorado, May, 161–169.
28. Humphrey, P. E., and Ethington, R. L. (1990). "Non-Destructive Testing Research at Oregon State University, Dept. of Forest Products," *Proc., Seventh International Nondestructive Testing of Wood Symposium*, Madison, Wisconsin, September 27–29, 1989, Conferences and Institutes, Washington State University, Pullman, Washington, 125–129.
29. Kayser, J. R., and Miller, E. D. (1992). "Accuracy in the Thickness Measurement of Corroded Bridge Steel Using an Ultrasonic Thickness Meter," *Proc., Nondestructive Evaluation of Civil Structures and Materials*, edited by B. A. Suprenant, J. L. Noland, and M. P. Schuller, University of Colorado, Boulder, Colorado, May, 241–246.
30. Kisters, F. H., and Kearney, F. W. (1991). "Evaluation of Civil Works Metal Structures," *Technical Report REMR-CS-31*, U.S. Army Construction Engineering Research Laboratory, Champaign, Illinois, January, 82 pp.
31. Lemaster, R. L., and Pugel, A. D. (1990). "Measurement of Density Profiles of Wood Product Materials Using Acoustic Emission," *Proc., Seventh International Nondestructive Testing of Wood Symposium*, Madison, Wisconsin, September 27–29, 1989, Conferences and Institutes, Washington State University, Pullman, Washington, 3–28.
32. Maji, A. K., and Shah, S. P. (1988). "Application of Acoustic Emission and Laser Holography to Study Microfracture in Concrete," Nondestructive Testing, *ACI Publication SP-112*, edited by

H. S. Lew, American Concrete Institute, Farmington Hills, Michigan, 83–109.

33. Maji, A. K., and Wang, J. L. (1992). "Inspection of Model Steel Bridge Components with Electronic Shearography," *Proc., Nondestructive Evaluation of Civil Structures and Materials*, edited by B. A. Suprenant, J. L. Noland, and M. P. Schuller, University of Colorado, Boulder, Colorado, May, 263–277.

34. Malhotra, V. M. (1984). "In Situ/Nondestructive Testing of Concrete—A Global Review," In Situ/Nondestructive Testing of Concrete, *ACI Publication SP-82*, edited by V. M. Malhotra, American Concrete Institute, Farmington Hills, Michigan, 1–16.

35. Mather, K. (1966). "Hardened Concrete-Petrographic Examination," Significance of Tests and Properties of Concrete and Concrete Making Materials, *ASTM STP 169A*, American Society for Testing and Materials, West Conshohocken, Pennsylvania, 125–143.

36. Matzkanin, G. A., Stephens, T. W., and Chalkley, A. J. (1992). "Development of a New Nondestructive Inspection System for Bridge Structures," *Proc., Nondestructive Evaluation of Civil Structures and Materials*, edited by B. A. Suprenant, J. L. Noland, and M. P. Schuller, University of Colorado, Boulder, Colorado, May, 22 pp.

37. Mazurek, D. F., Jordan, S. R., Palazzetti, D. J., and Robertson, G. S. (1992). "Damage Detectability in Bridge Structures by Vibrational Analysis," *Proc., Nondestructive Evaluation of Civil Structures and Materials*, edited by B. A. Suprenant, J. L. Noland, and M. P. Schuller, University of Colorado, Boulder, Colorado, May, 181–193.

38. McCuen, R. H., Aggour, M. S., and Ayyub, B. M. (1988). "Spacing for Accuracy in Ultrasonic Testing of Bridge Timber Piles." Journal of Structural Engineering, Vol. 114, No. 2, 2652–2668.

39. Mielenz, R. C. (1966). "Concrete Aggregate-Petrographic Examination," Significance of Tests and Properties of Concrete and Concrete Making Materials, *ASTM STP 169A*, American Society for Testing and Materials, West Conshohocken, Pennsylvania, 381–403.

40. Mlakar, P. F., Walker, R. E., Sullivan, B. R., and Chiarito, V. P. (1984). "Acoustic Emission Behavior of Concrete," In Situ/Nondestructive Testing of Concrete, *ACI Publication SP-82*, edited by V. M. Malhotra, American Concrete Institute, Farmington Hills, Michigan, 619–637.

41. Nast, T. E., Zoughi, R., and Nowak, P. S. (1992). "Preliminary Results of Microwave Reflectometry as a Nondestructive Tool for Studying Concrete Properties," *Proc., Nondestructive Evaluation of Civil Structures and Materials*, edited by B. A. Suprenant, J. L. Noland, and M. P. Schuller, University of Colorado, Boulder, Colorado, May, 339–344.

42. Nazarian, S., and Baker, M. (1992). "A New Nondestructive Testing Device for Comprehensive Pavement Evaluation," *Proc., Nondestructive Evaluation of Civil Structures and Materials*, edited by B. A. Suprenant, J. L. Noland, and M. P. Schuller, University of Colorado, Boulder, Colorado, May, 437–452.

43. O'Dell, D. A., Hansen, R. E., and McDowell, S. L. (1992). "Ultrasonic Testing of Fillet Welds," *Proc., Nondestructive Evaluation of Civil Structures and Materials*, edited by B. A. Suprenant, J. L. Noland, and M. P. Schuller, University of Colorado, Boulder, Colorado, May, 411–425.

44. Ohtsu, M. (1988). "Diagnostics of Cracks in Concrete Based on Acoustic Emission," Nondestructive Testing, *ACI Publication SP-112*, edited by H. S. Lew, American Concrete Institute, Farmington Hills, Michigan, 63–82.

45. Ohtsu, M. (1992). "Current NDE Developments of Concrete Structures and Materials in Japan," *Proc., Nondestructive Evaluation of Civil Structures and Materials*, edited by B. A. Suprenant, J. L. Noland, and M. P. Schuller, University of Colorado, Boulder, Colorado, May, 1–16.

46. O'Leary, P. N., Bagdasarian, D. A., and DeWolf, J. T. (1992). "Bridge Condition Assessment Using Signatures," *Proc., Nondestructive Evaluation of Civil Structures and Materials*, edited by B. A. Suprenant, J. L. Noland, and M. P. Schuller, University of Colorado, Boulder, Colorado, May, 171–179.

47. Olson, L. D., Sack, D. A., and Phelps, G. C. (1992). "Sonic NDE of Bridges and Other Concrete Structures," *Proc., Nondestructive Evaluation of Civil Structures and Materials*, edited by B. A. Suprenant, J. L. Noland, and M. P. Schuller, University of Colorado, Boulder, Colorado, May, 279–295.

48. Pastor, J. A., and Navas, A. (1992). "Evaluation of Earthquake Damaged Structures Using Non-Destructive Testing Methods," *Proc., Nondestructive Evaluation of Civil Structures and Ma-*

terials, edited by B. A. Suprenant, J. L. Noland, and M. P. Schuller, University of Colorado, Boulder, Colorado, May, 313–328.
49. Patton-Mallory, M., and De Groot, R. C. (1990). "Detecting Brown-Rot Decay in Southern Yellow Pine by Acousto-Ultrasonics," *Proc., Seventh International Nondestructive Testing of Wood Symposium*, Madison, Wisconsin, September 27–29, 1989, Conferences and Institutes, Washington State University, Pullman, Washington, 29–44.
50. Pellerin, R. F., DeGroot, R. C., and Esenther, G. R. (1985). "Nondestructive Stress Wave Measurements of Decay and Termite Attack in Experimental Wood Units," *Proc., Fifth Nondestructive Testing of Wood Symposium*, Pullman, Washington, September 9–11, Conferences and Institutes, Washington State University, Pullman, Washington, 319–352.
51. Pla, G., and Eberhard, M. O. (1992). "Applications of Imaging Technology to the Nondestructive Evaluation of Reinforced Concrete," *Proc., Nondestructive Evaluation of Civil Structures and Materials*, edited by B. A. Suprenant, J. L. Noland, and M. P. Schuller, University of Colorado, Boulder, Colorado, May, 99–112.
52. Platkin, D. E. (1991). "REMR Management Systems—Coastal/Shore Protection Structures: Condition Rating Procedures for Rubble Breakwaters and Jetties, Initial Report," *Technical Report REMR-OM-11*, U.S. Army Construction Engineering Research Laboratory, Champaign, Illinois, May, 38 pp.
53. Portala, J.-F., and Ciccotelli, J. (1990). "NDT Techniques for Evaluating Wood Characteristics," *Proc., Seventh International Nondestructive Testing of Wood Symposium*, Madison, Wisconsin, September 27–29, 1989, Conferences and Institutes, Washington State University, Pullman, Washington, 98–124.
54. Ross, R. J. (1985). "Stress Wave Propagation in Wood Products," *Proc., Fifth Nondestructive Testing of Wood Symposium*, Pullman, Washington, September 9–11, Conferences and Institutes, Washington State University, Pullman, Washington, 291–318.
55. Ross, R. J. (1992). "Nondestructive Testing of Wood," *Proc., Nondestructive Evaluation of Civil Structures and Materials*, edited by B. A. Suprenant, J. L. Noland, and M. P. Schuller, University of Colorado, Boulder, Colorado, May, 43–49.
56. Sansalone, M., and Carino, N. J. (1988). "Laboratory and Field Studies of the Impact-Echo Method for Flaw Detection in Concrete," Nondestructive Testing, *ACI Publication SP-112*, edited by H. S. Lew, American Concrete Institute, Farmington Hills, Michigan, 1–20.
57. Sansalone, M., and Poston, R. (1992). "Detecting Cracks in the Beams and Columns of a Post-Tensioned Parking Garage Structure Using the Impact-Echo Method," *Proc., Nondestructive Evaluation of Civil Structures and Materials*, edited by B. A. Suprenant, J. L. Noland, and M. P. Schuller, University of Colorado, Boulder, Colorado, May, 129–143.
58. Secretary of Transportation. (1991). "Highway Bridge Replacement and Rehabilitation Program, Tenth Report of the Secretary of Transportation to the United States Congress," U.S. Department of Transportation, Federal Highway Administration, Washington, D.C., 50 pp.
59. Shahawy, M. A., and Issa, M. (1992). "Load Testing of Transversely Prestressed Double Tee Bridges," *PCI Journal*, Precast/Prestressed Concrete Institute, Vol. 37, No. 2, March–April, 86–99.
60. She, A. C., Hjelmstad, K. D., and Huang, T. S. (1992). "Structural Damage Detection Using Stereo Camera Measurements," *Proc., Nondestructive Evaluation of Civil Structures and Materials*, edited by B. A. Suprenant, J. L. Noland, and M. P. Schuller, University of Colorado, Boulder, Colorado, May, 345–356.
61. St. John, D. A. (1990). "The Use of Large-Area Thin Sectioning in the Petrographic Examination of Concrete," Petrography Applied to Concrete and Concrete Aggregates, *ASTM STP 1061*, edited by B. Erlin and D. Stark, American Society for Testing and Materials, West Conshohocken, Pennsylvania, 55–70.
62. Sturrup, V. R., Vecchio, F. J., and Caratin, H. (1984). "Pulse Velocity as a Measure of Concrete Compressive Strength," In Situ/Nondestructive Testing of Concrete, *ACI Publication SP-82*, edited by V. M. Malhotra, American Concrete Institute, Farmington Hills, Michigan, 201–227.
63. Tamura, H., and Yoshida, M. (1984). "Nondestructive Method of Detecting Steel Corrosion in Concrete," In Situ/Nondestructive Testing of Concrete, *ACI Publication SP-82*, edited by V. M. Malhotra, American Concrete Institute, Farmington Hills, Michigan, 689–702.
64. Taylor, M. A. (1988). "Evaluation of Concrete Constituents Using Photon Radiation," Nonde-

structive Testing, *ACI Publication SP-112*, edited by H. S. Lew, American Concrete Institute, Farmington Hills, Michigan, 41–62.

65. Taylor, M. A. (1990). "Acoustic Emission Monitoring of Concrete Quality," *Proc., Non-Destructive Evaluation of Civil Structures and Materials*, edited by B. A. Suprenant, S. Sture, J. L. Noland, and M. P. Schuller, University of Colorado, Boulder, Colorado, October, 87–103.

66. Teodoru, G. V. (1988). "The Use of Simultaneous Nondestructive Tests to Predict the Compressive Strength of Concrete," Nondestructive Testing, *ACI Publication SP-112*, edited by H. S. Lew, American Concrete Institute, Farmington Hills, Michigan, 137–152.

67. Turkstra, C. J., Zoltanetzky, P., Jr., Lim, H. P., and Gordon, C. (1988). "A Statistical Study of the Correlation Between Field Penetration Strength and Field Cylinder Strength," Nondestructive Testing, *ACI Publication SP-112*, edited by H. S. Lew, American Concrete Institute, Farmington Hills, Michigan, 165–179.

68. Whiting, D. (1991). "Preliminary Tests of a New Surface Air-Flow Device for Rapid In Situ Indication of Concrete Permeability," *Presented at the 70th Annual Meeting, Paper No. 910071*, Transportation Research Board, Washington, D.C., January.

69. Whiting, D. A., and Matzkanin, G. A. (1992). "Field Testing of a Nondestructive Portable Permeability Indicator for Concrete," *Proc., Nondestructive Evaluation of Civil Structures and Materials*, edited by B. A. Suprenant, J. L. Noland, and M. P. Schuller, University of Colorado, Boulder, Colorado, May, 51–67.

70. Woodham, D. B. (1992). "Magnetically Induced Velocity Change to Measure Residual Stress," *Proc., Nondestructive Evaluation of Civil Structures and Materials*, edited by B. A. Suprenant, J. L. Noland, and M. P. Schuller, University of Colorado, Boulder, Colorado, May, 297–311.

71. Worthington, W. (1992). "Prestressed Concrete Pipe Inspection and Monitoring Methods," *Proc., Nondestructive Evaluation of Civil Structures and Materials*, edited by B. A. Suprenant, J. L. Noland, and M. P. Schuller, University of Colorado, Boulder, Colorado, May, 83–97.

72. Weil, G. J. (1992). "Non-Destructive Testing of Bridge, Highway and Airport Pavements," *Proc., Nondestructive Evaluation of Civil Structures and Materials*, edited by B. A. Suprenant, J. L. Noland, and M. P. Schuller, University of Colorado, Boulder, Colorado, May, 385–394.

73. Wynn, C. C., Fletcher, W. M., and Jones, W. D. (1992). "Nondestructive Evaluation of Hollow Clay Tile Walls," *Proc., Nondestructive Evaluation of Civil Structures and Materials*, edited by B. A. Suprenant, J. L. Noland, and M. P. Schuller, University of Colorado, Boulder, Colorado, May, 375–383.

74. Yun, C. H., Choi, K. R., Kim, S. Y., and Song, Y. C. (1988). "Comparative Evaluation of Nondestructive Test Methods for In-Place Strength Determination," Nondestructive Testing, *ACI Publication SP-112*, edited by H. S. Lew, American Concrete Institute, Farmington Hills, Michigan, 111–136.

4.0 EVALUATION PROCEDURES AND EVALUATION OF STRUCTURAL MATERIALS AND SYSTEMS

4.1 EVALUATION PROCEDURES

4.1.1 General

Evaluation, as used herein, is defined as the process of determining the structural adequacy of the building or component for its intended use. Evaluation by its nature implies the use of personal judgment by those qualified as experts. Because these experts must ultimately take the professional responsibility for the evaluation, the requirements for evaluation cannot be standardized. However, guidelines for the evaluation process are established in this Section for the purpose of reference.

4.1.2 Evaluation and Acceptance Criteria

Where practical, the engineer and client should agree on the criteria necessary for any evaluation. While most decisions will be based on experience, there may be resources available to guide the engineer.

4.1.3 Recognized Methods

Recognized reliable methods should be used for all evaluation procedures.

4.1.4 New Methods

The use of new or improved methods for evaluation is encouraged. When new methods are utilized in evaluation procedures, there should be reliable and sufficient checks of these new methods by use of recognized methods.

4.2 EVALUATION OF STRUCTURAL CONCRETE

Structural concrete includes unreinforced and reinforced concrete (both cast-in-place and precast concrete), prestressed concrete (pre-tensioned or post-tensioned), or combinations thereof.

4.2.1 Concrete

The function of concrete material in a structure is two-fold. First, the concrete functions as one component of the composite structural material that constitutes the load carrying element. Second, the concrete provides an overall protection against fire and environmental forces. Specifically, the concrete cover reduces or prevents corrosion of the embedded reinforcing or prestressing steel and provides for durability.

For concrete to function as a load carrying structural element, the following two coincidental characteristics are required: adequate compressive strength and adequate cross-sectional size which contributes to member stiffness. If the combination of these two characteristics in a member is not adequate, the member is unacceptable.

4.2.1.1 Causes of Concrete Deterioration

Causes of concrete deterioration include alkali-aggregate reactions; unsound cement; contaminated water and aggregates; sulphate attack; freezing and thawing; fatigue; and damage from accidents. Some of these causes are directly related to deterioration of the reinforcing steel and the prestressing steel embedded in the concrete.

Accident Damage. Concrete is a brittle material and when struck by a vehicle, for example, brittle fracture can result. In general, exposing concrete elements to vehicle traffic should be avoided or the elements should be protected against impact damage. When this is not done, damage is the result. Examples include columns and walls unprotected by guard rails in garages and loading areas.

Alkali-Aggregate Reaction (AAR). Reaction in concrete between alkalies from the cement in the pore water of the concrete and reactive silica in the aggregate is termed alkali-silica reaction (Struble and Diamond, 1986). This reaction can cause cracking and occasionally significant weakening of the concrete. There are believed to be three types of alkali-aggregate reactions; alkali-silica, alkali-silicate, and alkali-carbonate. The alkali-silica reaction is the most common (Hobbs, 1988). Alkali-silica reaction can cause substantial reductions in the engineering properties of concrete (Swamy and Al-Asali, 1986). At 0.1% expansion, compressive strength decreases about 12%, loss of flexural strength can be as much as 50%, and the elastic modulus is reduced approximately 20%. However, few structures have been replaced because of AAR.

Cement Soundness. Portland cement as a hydrated paste is the binder of most concretes. This binder governs most of the properties of concrete. Cement is available in several types, with and without air-entraining agents, to meet various construction requirements (such as early high strength) and for compatibility with certain types of aggregates. Cement can become contaminated—usually when it is transported or stored. Contamination by organic materials can result in a loss of the required properties of the concrete. Specifications for cement manufacture specify limitation in fineness and specific surface. Cements with a specific surface which is too low may produce concrete with poor workability and excessive bleeding. Decreased fineness increases water requirements while a greater fineness improves early strength development but reduces resistance to freezing and thawing. Because of differences in raw materials, cements with the same fineness and chemical analyses can have differences in strength, heat of hydration, production of laitance, bleeding tendency, and durability (U.S. Department of the Interior, 1981). As noted above, the alkalies in the cement can react with minerals in the aggregates resulting in cracking and expansion. Sulfates in groundwater and in soil, against which concrete is placed, can also react with chemicals in the cement resulting in disintegration of the concrete. This discussion indicates that what may be sound cement for one set of conditions may be unsound cement for a different set of circumstances. Cement must be selected so that it is compatible with site conditions, construction methods, required properties of the concrete, and the other ingredients of the concrete mixture.

Contaminated Water and Aggregates. The usual rule governing mixing water is that if it meets the requirements for drinking, it is satisfactory for making concrete. This is not an entirely satisfactory definition since water containing sugar or citrate flavoring are fit to drink but not for mixing concrete (McCoy, 1966). Mixing water can be contaminated by impurities such as sodium and potassium carbonates and bicarbonates, sodium chloride, sodium sulfate, calcium and magnesium bicarbonates, calcium chloride, iron salts, sodium iodate, phosphate arsenate, borate, sodium sulfide, hydrochloric and sulfuric acids, sodium hydroxide and salt and suspended particles. It can also be contaminated by algae and other organic materials. Minute concentrations are tolerable and limits on concentration of different impurities are available (McCoy, 1966).

Aggregates for concrete should generally be inert. However, aggregates can contain minerals which react unfavorably with the chemicals in the cement paste. The reactive components of aggregates include calcium, magnesium, silica and iron oxides; ferrous sulphides; glasses; calcium sulfate; zeolites; clay minerals; and dolomitic limestone (Hansen, 1966). As with unsound cement, some contaminated aggregates may be perfectly acceptable for specific uses, construction methods, and with the other components of the concrete mix, but the opposite may be true for different conditions.

Cracking, Scaling, Spalling, and Delamination. A crack is an incomplete separation of concrete into one or more parts of variable depth but visible at the surface. Plastic shrinkage cracks result from rapid drying of the concrete in its plastic state. Map cracking is a closely spaced network of cracks usually resulting from sulfate attack, frost attack and chemical reactions between the mineral aggregates and the cement paste. Drying shrinkage cracks result from the drying of restrained concrete after it has hardened. Settlement cracks can result from settlement of the form work as well as from settlement of the foundation. Structural cracks result from unreinforced or overstressed tension areas in the concrete. Corrosion induced cracks are caused by corrosion of reinforcing steel in the concrete (Section 3.7, Table 3.7.3, Manning, 1985). Much surface cracking seems to be due primarily to concrete volume changes caused by poor construction practices such as concrete that is too wet, placed too slowly, poorly finished, and improperly cured (Peterson, 1977).

Scaling is flaking away or disintegration of the surface mortar of concrete exposing and eventually loosening of coarse aggregate. Although a weak surface layer of mortar may result from poor finishing and curing practice, it is usually indicative of inadequate air entrainment (Section 3.7, Table 3.7.3, Manning, 1985). Scaling is not directly related to traffic (Peterson, 1977). It is often caused by freeze-thaw action in the absence of deicers.

Spalling is often caused by corrosion of reinforcing steel in the concrete and by impact. Local disintegration of a portion of the concrete surface (sometimes referred to as potholes) is also classified as spalling.

Delamination usually occurs over large subsurface areas of corroding reinforcing steel (rather than being localized over a few reinforcing steel bars) and results in a plane of separation within the concrete immediately over the plane of the reinforcing steel. If planes of reinforcing steel occur at several depths and the steel is corroding at each depth, then there may be more than one plane of delamination.

Degradation. Concrete can degrade when it comes into contact with both organic and inorganic acids. Organic materials include acetic acid, oxalic and dry carbonic acids, carbonic acid in water, lactic and tannic acid, and vegetable oils. Common inorganic acids include hydrochloric, sulfuric, nitric, and phosphoric acid. Deterioration can also be the result of environmental attack by wind and rain. Certain rain water can have a degrading affect on concrete.

Fatigue. Fatigue damage of reinforced concrete is dependent upon both the fatigue life of the reinforcing steel and of the concrete. Like all fatigue damage, the number of cycles of stress and the stress range are the most important parameters affecting fatigue damage. The failure of concrete under repeated loading results in progressive microcracking of the concrete. Since localized plastic deformations to blunt microcracks, such as occurs in metals, do not occur in concrete, concrete has no endurance limit similar to that for steel (Hawkins and Shah, 1982). The fatigue strength of concrete submerged in water is different from that of concrete in an air environment. Water trapped in the opening and closing of the cracks causes hydraulic fracturing. Submerged concrete has a shorter fatigue life than unsubmerged concrete. Fatigue of concrete can be of several different types or combinations of types. These include shear fatigue, tension fatigue and bond fatigue. Under static loads, microcracking associated with these failure mechanisms may be insignificant, but the repetition and range of stresses can result in a breakdown of resistance. Research on concrete indicates that fatigue damage of concrete may be more prevalent than realized (Lenschow, 1982). Also, the higher degree of utilization of the capacity of materials, which increases stress variations and brings the service stress range closer to the failure stress, will result in a greater amount of concrete fatigue damage. Much of the spalling of concrete from the underside of concrete parking garage decks and from reinforced concrete beams and girders subjected to cyclical stresses is suspected to be fatigue related.

Fire. Fire can damage reinforced concrete in several ways. Surface defects such as spalling and cracking can occur as a result of internal pore pressure build-up, induced thermal stresses caused by severe temperature gradient, or from differences in thermal expansion characteristics of two unlike materials. Over time, with increasing fire exposure, both concrete and steel will reach temperatures that result in loss of strength. Cement paste decomposes and progressively begins to lose strength at about 300°C.

Freezing and Thawing. Disintegration and cracking of concrete can be caused by the disruptive action of freezing and thawing. Dry concrete, with or without entrained air sustains no damaging effects from freezing and thawing. It is the water that is available in the capillary system that freezes and expands with resulting concrete disintegration. In general, the problem is controlled by air entrainment that provides relief for pressures developed by free water as it freezes and expands.

Sulphate Attack. Sulphates in ground water (or mixing water), or in soils against which concrete will be in contact, can react with the cement paste in concrete and result in disintegration of the concrete. The problem is fairly common in arid regions and in exposures to seawater (Reading, 1975; Mehta and Polivka, 1975). Sulphates of sodium and magnesium can react with the cementing constituents of the concrete. Depending upon the composition of the cement, these chemical reactions may be accompanied by large volume changes in the concrete.

4.2.1.2 Properties and Physical Conditions

As a structural material, the properties and physical conditions of concrete that are summarized in Tables 4.2.1(a) and 4.2.1(b) should be considered in evaluating the acceptability of existing concrete.

4.2.2 Reinforcing Steel, Pre-Tensioning Steel, Post-Tensioning Steel and Cable Steel

The function of reinforcing steel, pre-tensioning steel, post-tensioning steel and cables in a concrete structure is to carry and transmit either compressive or tensile stresses. Not only must the properties and physical conditions of the steel be determined to evaluate this stress carrying ability, but the means of transmitting and distributing the stresses to the concrete structure must also be determined.

4.2.2.1 Causes of Reinforcing Steel Deterioration

Reinforcing steel is a component of reinforced concrete. The primary cause of deterioration of reinforcing steel is corrosion. A secondary cause is fatigue cracking. Both of these types of deterioration trigger deterioration in the surrounding concrete. While both materials have separate deterioration mechanisms, the deterioration of one affects the other.

Corrosion. Reinforcing steel is surrounded by concrete which provides a passive environment for the steel. Fresh concrete is highly alkaline with a pH in the range of 12 to 13. In this alkaline environment, the thin film of iron oxide that is usually present at the surface of the metal, is stabilized into a protective passive film (Borgard et al., 1990). This passive film protects the reinforcing steel against corrosion. In addition, concrete of low water-cement ratio, and well cured, has low permeability which minimizes penetration of corrosion inducing factors, especially oxygen, chlorides and carbon dioxide. The low permeability also increases the electrical resistivity of the concrete which assists in reducing the rate of corrosion by retarding the flow of electrical currents that accompany electrochemical corrosion (Verbeck, 1975).

Corrosion of reinforcing steel is the electrochemical degradation of the steel. It occurs in the presence of oxygen and moisture whenever the protective passive film is destroyed by either carbonization or by the presence of more than the threshold concentration of chloride ions at the steel surface. Carbonation is the result of the reaction of carbon dioxide and other acidic gases in the air and the alkaline constituents of the cement paste. As a result the alkalinity of the concrete (pH) is reduced and the steel loses its protective film. Chloride ions can also penetrate the protective passive film, with an accompanying shift in steel potential and subject the steel to the action of oxygen and moisture and resulting corrosion (Locke, 1986).

Once the protective oxide film has been impaired, and in the presence of oxygen and moisture, corrosion of the steel will proceed. There are two mechanisms of steel corrosion; one electrochemical/physical and the other galvanic/physical. In marine environments or where salt is used as a deicing agent as in parking garages or areas open to vehicular traffic, chlorides will permeate the concrete by absorption, capillarity, or diffusion through cracks (Mehta, 1977). Chloride ions depassifies steel which results in electrochemical corrosion cell and, with oxygen and

TABLE 4.2.1(a) (1 of 2). Evaluation of Properties of Concrete

References: See numerical references from noted tables in Section 3.7 / Chemical and Physical Properties \ Evaluation Procedure	Table 3.7.1 4,11, 17,18 22	Table 3.7.2 8,23	Table 3.7.2 4,17 20,22	Table 3.7.2 15	Table 3.7.1 2,4,5, 9,22,	Table 3.7.1 4,8, 10,16 17,21 22	Table 3.7.2 4,16, 18,22	Table 3.7.2 15, Table 3.7.3 3,8	Table 3.7.1 4,11, 17,18 20,22	Table 3.7.1 4,11, 13,17 19,21 22	Table 3.7.2 1,10 15,24 25,26	Table 3.7.2 2,8,9 15,17	Table 3.7.2 8,14, 21		
	Air Content Test	Cast-In Place Procedure (Pull-Out Test)	Cement Content Test	Electrical Resistance Measurements	Flexural Tests	Freeze-Thaw Test	Gamma Radiography	Nuclear Moisture Meter	Petrographic Analysis	Test on Concrete Cores	Ultrasonic Pulse	Windsor Probe	Pull-Off Testing		
Acidity									•	•					
Air Content	•						•		•	•					
Cement Content			•						•	•					
Chemical Content									•	•					
Chloride Content									•	•					
Compressive Strength		•								•	•	•	•		
Creep										•					
Density							•			•					
Elongation										•					
Modulus of Elasticity											•	•			
Modulus of Rupture					•										
Moisture Content				•				•							
Permeability										•					
Proportions of Aggregate										•					
Pull Out Strength		•													
Resistance to Freezing and Thawing						•				•					
Soundness									•	•					
Splitting Tensile Strength										•					

TABLE 4.2.1(a) (2 of 2). Evaluation of Properties of Concrete

References: See numerical references from noted tables in Section 3.7	Table 3.7.1 4,11, 17,18 22	Table 3.7.2 8,23	Table 3.7.2 4,17 20,22	Table 3.7.2 15	Table 3.7.1 2,4,5, 9,22	Table 3.7.1 4,8, 10,16 17,21 22	Table 3.7.2 4,16, 18,22	Table 3.7.2 15, Table 3.7.3 3,8	Table 3.7.1 4,11, 17,18 20,22	Table 3.7.1 4,11,1 3,1719 ,2122	Table 3.7.2 1,10 15,24 25,26	Table 3.7.2 2,8,9 15,17	Table 3.7.2 8,14, 21		
Evaluation Procedure / Chemical and Physical Properties	Air Content Test	Cast-In Place Procedure (Pull-Out Test)	Cement Content Test	Electrical Resistance Measurements	Flexural Tests	Freeze-Thaw Test	Gamma Radiography	Nuclear Moisture Meter	Petrographic Analysis	Test on Concrete Cores	Ultrasonic Pulse	Windsor Probe	Pull-Off Testing		
Tensile Strength		●			●					●					
Uniformity of Mix									●						
Water-Cement Ratio								●	●						

TABLE 4.2.1(b) (1 of 2). Evaluation of Physical Conditions of Concrete

References: See numerical references from noted tables in Section 3.7	Table 3.7.3 8,10 21,27	Table 3.7.4 24	Table 3.7.3 8,9 16	Table 3.7.3 11,17 18,19 30,31	Table 3.7.13 ALL REF.	Table 3.7.3 8,9, 27	Table 3.7.1 4,11, 17,18 20,22	Table 3.7.3 1,16	Table 3.7.3 5,7, 11,13 17,18	Table 3.7.2 3,7,8 11,15 16	Table 3.7.1 4,11, 13,17 19,21 22	Table 3.7.3 16,22 23,26 28,32	Table 3.7.3 1,16	Table 3.7.2 2,8,9 15,17
Chemical and Physical Properties \ Evaluation Procedure	Acoustic Emission\Impact	Fiber Optics	Gamma Radiography	Infrared	Load Testing	Pachometer	Petrographic Analysis	Physical Measurement	Radar	Schmidt Rebound Hammer	Tests on Concrete Cores	Ultrasonic Pulse	Visual Examination	Windsor Probe
Alkali-Carbonate Reaction							●							
Alkali-Silica Reaction														
Bleeding Channels													●	
Cement Aggregate Reaction							●							
Chemical Deterioration													●	
Chloride Attack							●						●	
Contaminated Aggregate							●							
Contaminated Mixing Water							●							
Cracking	●							●				●	●	
Cross-Sectional Properties			●					●				●		
Delamination	●		●	●					●			●	●	
Deterioration			●	●			●				●	●	●	
Discoloration							●						●	
Disintegration			●	●			●					●	●	
Distortion													●	
Efflorescence				●			●						●	
Erosion							●						●	
Freeze-Thaw Damage							●						●	

TABLE 4.2.1(b) (2 of 2). Evaluation of Physical Conditions of Concrete

References: See numerical references from noted tables in Section 3.7 / Chemical and Physical Properties \ Evaluation Procedure	Acoustic Emission/Impact (Table 3.7.3 8,10 21,27)	Fiber Optics (Table 3.7.4 24)	Gamma Radiography (Table 3.7.3 8,9 16)	Infrared (Table 3.7.3 11,17 18,19 30,31)	Load Testing (Table 3.7.13 ALL REF.)	Pachometer (Table 3.7.3 8,9, 27)	Petrographic Analysis (Table 3.7.1 4,11, 17,18 20,22)	Physical Measurement (Table 3.7.3 1,16)	Radar (Table 3.7.3 5,7, 11,13 17,18)	Schmidt Rebound Hammer (Table 3.7.2 3,7,8 11,15 16)	Tests on Concrete Cores (Table 3.7.1 4,11, 13,17 19,21 22)	Ultrasonic Pulse (Table 3.7.3 16,22 23,26 28,32)	Visual Examination (Table 3.7.3 1,16)	Windsor Probe (Table 3.7.2 2,8,9 15,17)
Honeycomb			•	•			•				•	•	•	
Leaching							•						•	
Popouts							•						•	
Scaling							•						•	
Spalling			•	•							•	•	•	
Stratification							•						•	
Structural Performance	•				•								•	
Sulphate Attack							•						•	
Uniformity of Concrete			•				•			•		•	•	•
Unsound Cement							•							
Unsound Concrete							•				•	•	•	•

moisture, accelerates attack on the reinforcing steel. Corrosion products from the attacked steel create expansive forces (physical) within the concrete, causing it to spall and crumble. As the steel is further exposed to oxygen, CO_2 and Cl^-, it corrodes more and increases deterioration of the concrete in a progressive manner.

Likewise, an imbalance of temperature, oxygen concentration, atmospheric pressure, stray current or other variables produce galvanic corrosion cells. The resulting products of corrosion cause deterioration of the concrete. Conditions for galvanic and physical corrosion ideally exist when reinforcing steel embedded in concrete is partially exposed to seawater.

Fatigue. Fatigue failures of reinforcing steel in tension can generally be attributed to the magnitude of stress range in the reinforcement and to stress raisers provided by deformation lugs (Helgason et al., 1976). There are other contributing factors such as stress concentration at critical points such as at bar anchorages, splices, and bar bends; and friction (fretting) between the reinforcing steel and the concrete at a cracked section (Naaman, 1989). Current design specifications provided by ACI Committee 343 and AASHTO address fatigue stress limits in reinforcing steel under the Strength Design Method and under Load Factor Design. Both specifications limit the stress range of straight tensile reinforcement. It is concluded that the current design specifications are conservative at least for straight tensile reinforcement.

Most fatigue failures in reinforcing steel can be associated with the stress concentration at critical points. These include butt-welded reinforcement (Walls et al., 1965), splice couplings and sleeves (Tassi and Magyari, 1982; Bennett, 1982; Kordina and Quast, 1982), notch effect of the deformations, and the ratio of curvature of bent bars (Nurnberger, 1982).

Since allowable design stresses for fatigue of reinforcing steel are generally conservative, fatigue failure can usually be associated with corrosion of the reinforcing steel, which results in an increase in the magnitude of the stress, at the critical points of high stress.

4.2.2.2 Causes of Prestressing Steel Deterioration

Like reinforcing steel, prestressing steel deterioration can result from corrosion and fatigue. Because of the high stresses associated with prestressing steel, hydrogen embrittlement and stress corrosion are also possible causes of deterioration. When prestressing steel is encased in conduits for post-tensioning, the lack of grouting of these conduits, and also inadequate grouting, may contribute to prestressing steel deterioration.

Corrosion. The discussion on corrosion of reinforcing steel applies directly to prestressing steel. Corrosion of prestressing steel appears to be more highly sensitive to permeable concrete associated with a high water cement ratio, honeycomb and large bleed holes (Gerwick, 1975). Also, stray electric current appears to affect the corrosion process for prestressing steel, at least for prestressed concrete exposed to sea water, more than for reinforcing steel (Cornet et al., 1980). Corrosion of prestressing steel is usually associated with catastrophic failure (Rothman and Price, 1986). Anchorages and couplings for prestressing steel usually have less protective concrete cover and are accordingly more susceptible to corrosion.

Fatigue. The discussion on fatigue of reinforcing steel applies directly to prestressing steel. Points of critical stress for prestressed members occur at changes in direction of the prestressing steel and at the anchorages. In partially prestressed concrete beams, the concrete section is generally uncracked under dead load with cracking initiating under the first application of live load. Subsequent applications of live loads will lead to cracks opening at the decompression stress which is lower than the cracking stress.

The location of the neutral axis will shift upward, leading to a higher rate of increase in the steel stress as well as the stress in the extreme fiber of the concrete. These repetitive changes in stress create fatigue damage in the corresponding materials, reduce bond properties at the interface between the prestressing steel and the concrete, and lead to substantial increases in crack widths and deflections under service loads (Naaman, 1982). Consequently, partially prestressed concrete members generally have shorter fatigue lives than fully prestressed concrete members.

Hydrogen Embrittlement. Hydrogen embrittlement is a failure mechanism of metal resulting from the entry of hydrogen into the metal, which reduces its ability to deform plastically (Sheinker and Wood, 1972). It is quite often identified as one of the two mechanisms of stress corrosion (the other mechanism of stress corrosion is identified as the active path corrosion type) but is usually treated as a separate subject. In order for either hydrogen embrittlement or stress corrosion cracking to occur the following conditions are necessary (Sheinker and Wood, 1972):

1. A susceptible metal. A metal may be susceptible to hydrogen embrittlement or stress corrosion cracking in only a few specific environments.
2. A specific environment. A particular environment may induce cracking in only certain metals.
3. A tensile stress. The tensile stress usually must exceed a certain level depending upon the particular metal-environmental couple.

ASTM A143, 1994, defines embrittlement as the loss or partial loss of ductility in a steel often associated with strain aging which proceeds slowly at room temperatures or more rapidly as the aging temperature is raised. Accordingly, galvanized prestressing steel is more susceptible to hydrogen embrittlement than ungalvanized prestressing steel because of the galvanizing temperature of approximately 850°F (455°C). Hydrogen embrittlement can result from atomic hydrogen being absorbed by the steel during the steel manufacturing and fabrication processes. It can also result from hydrogen introduced during service or environmental exposure such as during welding (Raymond, 1988).

One reason for separating the phenomenon of hydrogen embrittlement from stress corrosion is that the two types of cracking failures respond differently to environmental variables (Fontana and Greene, 1978). For example, cathodic protection is an effective means of preventing stress corrosion cracking whereas it rapidly accelerates hydrogen embrittlement effects.

The mechanism of hydrogen embrittlement can be characterized by three separate processes (Scully and Moran, 1988).

Inadequate Grouting. Prestressing steel for post-tensioned structures is installed in conduits to prevent bonding with the concrete when it is placed. After post-tensioning, these conduits are either filled with an organic substance to prevent bond or filled with a cement grout to provide a passive protective environment for the prestressing steel and to bond the prestressing steel to the concrete. Organic grouted conduits are termed unbonded and cement grouted conduits are termed bonded conduits. Without getting into the pros and cons of unbonded prestressing steel versus bonded prestressing steel, it is noted that both types of construction have a similar problem. Whether the conduits are filled with an organic substance or with a portland cement grout, inadequate filling of the conduit can occur. The resulting voids permit water leakage into the ducts. This water can contain impurities leading to a serious corrosion (Greiss and Naus, 1980) or the water can freeze with the resultant expansion cracking the concrete—usually at locations of minimum cover.

1. Adsorption—The process in which atomic hydrogen is accumulated and adheres to the surface of the metal. This process does not occur if the source of atomic hydrogen is within the metal.
2. Absorption—The process in which the atomic hydrogen is assimilated and incorporated into the metal. This process does not occur if the source of atomic hydrogen is within the metal.
3. Transportation—The process in which the atomic hydrogen moves to the embrittlement site.

Each of these three processes can occur in several different ways, some of which are not clearly understood at this time and research is still in progress to define the mechanism of hydrogen embrittlement in further detail.

Stress Corrosion. The characteristics of stress corrosion cracking may be summarized as follows (Brown, 1972):

1. Tensile stress is required. This stress may be supplied by service loads, cold work, mismatch in fit-up, heat treatment, and by the wedging action of corrosion products.
2. Only alloys are susceptible, not pure metals, though there may be a few exceptions to this rule.
3. Generally only a few chemical species in the environment are effective in causing stress corrosion cracking of a given alloy.
4. The species responsible for stress corrosion cracking in general need not be present either in large quantities or in high concentrations.
5. With some alloy/corrodent combinations, such as titanium and crystalline sodium chloride, or austenitic stainless steel and chloride solutions, temperatures substantially above room temperature may be required to activate some process essential to stress corrosion cracking.
6. An alloy is usually almost inert to the environment which causes stress corrosion cracking.
7. Stress corrosion cracks are always macroscopically brittle in appearance, even in alloys which are very tough in purely mechanical fracture tests. (Shear lips may occur in conjunction with stress corrosion cracks, but these shear lips are not part of the stress corrosion process. As a corollary, there does not appear to be a stress

corrosion fracture analog to the full shear slant fracture of purely mechanical origin.)
8. Microscopically, the fracture mode in stress corrosion cracking is usually different from the fracture mode in plane strain fractures of the same alloy.
9. There appears to be a threshold stress below which stress corrosion does not occur, at least in some systems.

The mechanism involved in stress corrosion cracking is not fully understood (Fontana and Greene, 1978). Corrosion plays an important part in the initiation of stress corrosion cracks. A corrosion pit first forms and acts as a stress raiser. There is a porous cap to the corrosion pit which impedes exchange between the corrodent within the pit and the bulk corrosion outside the pit. The essential feature of the corrosion pit is to provide a mechanism for altering the solution chemistry locally to one favorable for stress corrosion cracking. With the passage of time, cracks emanate from the pit downward into the metal. These cracks can grow to lengths sufficient for the remaining metal to fail in a purely mechanical brittle fracture (Brown, 1972). As noted under Hydrogen Embrittlement, this mechanism is termed active path corrosion.

4.2.2.3 Causes of Cable Deterioration

Cables are used for supporting roofs and for suspending floors from above. Most of the discussion on causes of reinforcing steel and prestressing steel deterioration applies to cable deterioration.

However, there are sufficient differences in the types of forces that cause deterioration, in the locations of points of deterioration, and in the details of construction, to warrant a separate discussion on cable deterioration.

Corrosion. Most cable supported structures are long-span. Because of their long span nature they are specially sensitive to aerodynamic forces due to wind and earthquakes, temperature changes, and the large movements associated with long-span structures. The magnitude and continuity of movements to which these structures are subjected to can aggravate the problems of corrosion by breaking down the cable protective systems through stress reversal and movement. All cables are made up of wires or ropes that have to be terminated by some sort of mechanical detail such as a socket or anchorage. The wires and cable ropes are usually splayed at the socket or anchorage to facilitate termination. The very nature of cable and rope construction means that there are ready made internal conduits that permit the free movement of moisture and air. All cables are protected by external coverings that in the past have sometimes proved unsuccessful in keeping moisture and air from penetration into the cable. Polyethylene ducts, epoxy coated cables, paint, galvanized wire, internally sealed cables, wire wrapping, combinations of these, and all manner of protective coatings and systems have been tried—some with more success than others—but many are problem prone.

While the processes and mechanisms of corrosion for cables are the same as for reinforcing steel and prestressing steel, for the reasons noted, cables are especially susceptible to corrosion at specific locations. These locations are defined by the possibility of accumulation of water and include the socket ends of cable and ropes, saddles and anchorages of cables, and clamp locations in suspender ropes that introduce details where water can accumulate. Accordingly, cables can have more serious corrosion problems than reinforced concrete or prestressed concrete construction, but these problems are usually isolated to specific details of the construction.

Fatigue. The aerodynamic nature of the forces associated with cable supported structures means that the cables are subjected to a greater range of stresses including reversal than other types of structures. The magnitude of the stress range is the primary contributing factor to fatigue failure. While present design practices are to limit the stresses in cables so that fatigue lives are acceptable, the associated problem of corrosion due to lack of maintenance results in increases in the stress range and resulting fatigue failures.

Hydrogen Embrittlement. Like prestressing steel, and for the same reasons, hydrogen embrittlement of wires in cables can be a problem. The extent of the problem, if any, is not known. While there is considerable literature documenting wire breaks in cable, the reasons for the breaks are simply attributed to corrosion. Hydrogen embrittlement failure determination, while usually associated with corrosion, requires specific tests to determine if embrittlement is part of the problem. The literature available does not report any such tests. However, conditions can be right for hydrogen embrittlement failures and it is suspected that the lack of reports on the subject is that the problem has not been investigated sufficiently.

Stress Corrosion. Stress corrosion does not appear to be a problem for wires in cables. However, like hydrogen embrittlement, conditions that exist for cables suggest that some of the wire breaks reported in the literature are the result of stress corrosion. Again, testing to verify the reasons for specific wire breaks is lacking and the extent of the problem is unknown.

4.2.2.4 Properties and Physical Conditions

As structural materials, the properties and physical conditions of reinforcing steel, pre-tensioning steel, post-tensioned steel, and cables that are summarized in Tables 4.2.2(a) and 4.2.2(b) should be considered in evaluating the acceptability of existing reinforcing steel, pre-tensioning steel, post-tensioning steel and cable steel.

4.2.3 Connections

Reference is made to the appropriate material section for evaluation of the properties and physical conditions of connections.

4.2.4 Connectors

Reference is made to the appropriate section for connectors for metals for the evaluation of the assessment of the properties and physical conditions of connectors.

4.3 EVALUATION OF METALS

Metals include all grades of hot-rolled and cold-formed structural steel, stainless steel, cast iron and wrought iron, parallel and helical steel wire structural strand, steel structural rope, steel rods, aluminum alloy, and other metals.

4.3.1 Structural Steel

The function of structural steel in a structure is to provide the strength and stiffness to carry the loads. This function requires that the chemical and physical properties of the steel remain adequate for the expected loading conditions. The chemical and physical properties of steel can be changed by deterioration, physical damage, fatigue, excessive loading conditions, exposure to fire, or extremely low temperatures. Loss of effective compressive bracing is usually associated with physical damage or alterations of the bracing components and systems. The fire retardant protective system must continue to function to maintain the required fire rating of the steel. The fire retardant protective system is usually changed by physical damage or alterations.

4.3.1.1 Causes of Structural Steel Deterioration

Deterioration of structural steel can take place by atmospheric corrosion; fatigue; exposure to soil, water, corrosive marine and industrial atmospheres or some combination of these exposures; leakage from stressed steel tanks, or pipes; and stress corrosion. Fracture, or a break in the mechanical continuity, is the end result of deterioration. These several causes are interrelated in the sense that more than one mechanism may be prevalent in the deterioration process at the same time. An additional cause of structural steel deterioration is damage from accidents.

Corrosion. Atmospheric corrosion (to distinguish the mechanism from crevice corrosion, pitting corrosion, stress corrosion, etc.) is generally an electrochemical process (Mattsson, 1982). This process takes place in small local galvanic cells which are similar to small batteries with anodes and cathodes acting as the electrodes. These cells operate only in the presence of an electrolyte which means that corrosion occurs only when the surface is wet. This moisture film will contain various species deposited from the atmosphere as well as those originating from the corroding metal. Oxygen will be readily absorbed from the air so that at least the outer region of the thin water film on the metal surface may be considered saturated with oxygen. Sulphur oxides, nitrogen oxides and chlorides will also be absorbed from the atmosphere. The concentrations of these various species absorbed from the atmosphere in the moisture film serves as the electrolyte. The difference in potential results from the fact that different metal surfaces possess different solutions or electrode potentials. This difference in potential results in flow of electric current and causes local attack to begin. The corrosion rate (usually considered as rate of metal loss or corrosion cement density) depends on many factors including the amount and composition of the electrolyte on the surface, the relative area of the anodes and cathodes, and the potential difference between them.

Steels with improved resistance to corrosion are called weathering steel including stainless steels and high alloy steels. Rust developing on weathering steels is more protective than rust on unalloyed or carbon steel. The penetration of rust approaches a nearly constant value after some years. This layer of rust then offers a protective coating known as a patina. However, when the atmosphere is highly polluted with sulphur dioxide and under climatic conditions with long wet periods, weathering steel behaves much like unalloyed carbon steels. The same is true

TABLE 4.2.2(a). Evaluation of Properties of Reinforcing Steel, Pre-Tensioning Steel, and Post-Tensioning Steel

References: See Section 3.7 and Table 3.7.4	17,21	1,3 23,24	2,4,5 7-16 18-20	6,22
Evaluation Procedure / Chemical and Physical Properties	Chemical Analysis	Coating Tests	Material Specification	Tension Tests
Bend Test			●	
Breaking Strength				●
Carbon Content	●			
Chemical Composition	●	●		
Coating Properties				
Adhesion		●		
Continuity		●		
Thickness		●		
Deformation Requirements			●	
Elongation				●
Reduction of Area				●
Strength of Connections			●	
Tensile Strength				●
Weld Shear Strength			●	
Yield Strength				●

TABLE 4.2.2(b). Evaluation of Physical Conditions of Reinforcing Steel, Pre-Tensioning Steel, and Post-Tensioning Steel

Physical Condition \ Evaluation Procedure	Cover Meter	Gamma Radiography	Probe Holes	Physical Measurement	Radar	Visual Examination
References: See Section 3.7 and Table 3.7.3	8, 9, 27	8, 9, 16			5,7,9, 11, 13, 17, 18, 20, 21, 29	1, 16
Anchorage	●		●	●		●
Corrosion						●
Cover	●	●	●	●	●	
Cross-sectional Properties				●		
Exposure						●
Location	●	●	●	●	●	
Shape	●		●	●		

for weathering steels in marine atmospheres, especially in sheltered positions, in tunnels, on soil, on concrete and in immersion situations.

Atmospheric corrosion is often the initial stage in deterioration of steel, weakening the member and making it susceptible to mechanical failure or leakage, fatigue or brittle cracking and stress corrosion.

Fatigue. The American Society of Testing and Materials (ASTM Committee on Terminology, 1990) defines fatigue as "the process of progressive localized permanent structural change occurring in a material subjected to conditions that produce fluctuating stresses and strains at some point or points and that may culminate in cracks or complete fracture after a sufficient number of fluctuations." Other significant ASTM definitions relating to fatigue are:

1. Fatigue Life, N—The number of cycles of stress or strain of a specified character that a given specimen sustains before failure of a specified nature occurs.
2. Fatigue Limit, S_f—The limiting value of the median fatigue strength as N becomes very large. Also, the point on the stress-strain curve below which no fatigue can be obtained regardless of number of cycles.
3. S-N Diagram—A plot of stress against the number of cycles to failure.

Some of the factors known to affect the fatigue strength (Munse, 1964) include:

1. Stress concentrations,
2. Number of cycles or repetitions of load,
3. Magnitude, nature and range of applied stress,
4. Material properties,
5. Size effects,
6. Surface finish,
7. Rate of load application,
8. Residual stresses,
9. Rest periods, and
10. Temperature.

Present design procedures treat all of these factors adequately so that theoretically there should not be a fatigue problem. This is not true for older, existing structures which may have been designed and constructed without adequate consideration having been given to fatigue. Even with new buildings, details may be used that have not been properly evaluated for fatigue life and fatigue limit and flaws may exist in materials, or be introduced during the fabrication and construction processes, that create stress concentrations above values anticipated.

The most important parameters affecting fatigue are the number of load repetitions and magnitude and range of stress due to these repetitions. For specific materials and design details, the Fatigue Life, N, and the Fatigue Limit, S_f, are determined in the laboratory to develop an S-N diagram for the specific material and detail. Based on the results of these tests, design specifications limit allowable stresses or load factors for members especially susceptible to fatigue, termed fracture critical members, and for members with specific fatigue susceptible details. Accordingly, most fatigue failures develop because of stress concentrations due to the nature of specific details, or due to defects or residual stresses in the material or component that result from the manufacturing, fabrication and construction processes.

Fatigue cracking and subsequent fracture occurs in tension zones of structural members and may be displacement induced (the geometry of the detail allows excessive distortion with resulting stress concentration), induced by initial defects or geometry, or induced by aeroelastic action. Atmospheric corrosion can play a part by reducing the cross-sections of the resisting elements, thereby increasing the magnitude of stress over a given range. The combination of unanticipated levels of stress over a stress range in a tension member subjected to repeated cycles of load results in fatigue cracking and fracture.

Stress Corrosion. Fracture due to stress corrosion is not a common problem in buildings and is usually associated with specific connection details as eye bars and pin connections. A rare example is the Berlin Congress Hall Roof (Isecke, 1982). Like fatigue cracking and fracture, stress corrosion is usually initiated by flaws or defects in a detail which generate high concentrations of stress along with atmospheric corrosion which reduces member section and also increases stresses. With these conditions, alloy steel tensile members in the right environment can fracture from stress corrosion.

Damage From Accidents. While not only an obvious cause of structure steel deterioration, damage from accidents is not uncommon. They usually result from structural members being struck by vehicles in building garages and loading areas, or by ships berthing at wharves supporting buildings. Structural members can also be damaged by fire resulting from vehicle accidents.

4.3.1.2 Properties and Physical Conditions

As a structural material, the properties and physical conditions of steel that are summarized in Tables

4.3.1(a) and 4.3.1(b) should be considered in evaluating the acceptability of existing structural steel members and connections.

4.3.2 Cast Iron and Wrought Iron

All applicable provisions of Sections 4.3.1 apply. Cast and wrought iron materials must be carefully analyzed because of known tendencies to behave poorly under high tensile and shear loadings.

4.3.3 Aluminum Alloy

All applicable provisions of Section 4.3.1 apply.

4.3.4 Connections

Connections include the component metal shapes, wire ropes, wire strands, and rods used to attach one structural element to another. All applicable provisions of Section 4.3.1 apply. For wire ropes, wire strands and rods, the applicable provisions of Section 4.3.5 apply.

4.3.5 Connectors

Connectors are the means of making connections. They include welds, rivets, bolts and rods, studs, and clamps. They are required to transmit forces between the elements that are being connected. As structural materials, the properties and physical conditions of connectors that are summarized in Tables 4.3.5(a) and 4.3.5(b) should be considered in evaluating the acceptability of existing connectors.

4.4 EVALUATION OF MASONRY

Masonry includes all non-load-bearing walls, load-bearing walls, beams, columns, piers, panels, curtain walls, floor arches, arches, and shells made from both non-reinforced and reinforced manufactured masonry units and natural building stone masonry units. These units are usually bedded or bonded together by mortar and grout using masonry ties, anchors and joint reinforcement as connectors. Masonry can also be used compositely with concrete, metals and wood. Reference is made to Sections 4.2, 4.3 and 4.5 respectively for evaluation of these materials.

Masonry is an assemblage of two or more components (see Section 3.4.2). Masonry assemblages generally provide multiple functions. Because of their space enclosing nature, they not only serve as structural elements to carry building loads, but also provide protection from the environment. Accordingly, in addition to evaluation of the properties of masonry, evaluation of the physical conditions can be approximately divided into those conditions that are influenced by the environment and those conditions influenced by the structure. Of course there is some overlap as can be expected. In addition to environmental and structural influences on the physical conditions of masonry, related architectural features of the construction that are not part of the masonry assemblage may also directly influence the physical conditions. Investigation must explore how the total assemblage fits together and how the various components are tied or held together.

Reference is made to Sections 4.2 and 4.3 for the evaluation of reinforcing steel bars, masonry ties, anchors and joint reinforcement. For strength of anchors and ties in masonry, see References Section 3.7, Table 3.7.9, ASTM E488 and ASTM E754. For the evaluation of concrete, metal and wood used compositely with masonry, reference is made to Sections 4.2, 4.3 and 4.5 respectively.

4.4.1 Causes of Masonry Deterioration

Masonry includes not only manufactured units as brick, but also natural building stone. Very rarely are the manufactured or natural properties of masonry causes for deterioration. The physical conditions that masonry is subjected to during its service life are the primary causes of deterioration. For convenience, these conditions have been grouped into environmental attack, structural incompatibility, inadequate architectural features and mortar and grout deterioration (Grimm, 1985b; Raths, 1985). For causes of deterioration of reinforcing steel in reinforced masonry, reference is made to the previous sections.

Environmental Attack. Environmental attack that cause deterioration of masonry includes deicing salts, freezing and thawing, moisture and weathering. Some of these conditions occur together. Driving rain cannot only introduce salts into the masonry, which results in salt fretting of stone (a pattern of erosion caused by salts, usually from the salting of sidewalks and pavements), but also can contribute to water penetration and erosion of the surface, edges and corners of stone masonry. Moisture or water penetration into the interconnected pores (measured by the permeability of the masonry) and through cracks and other physical openings can result in efflorescence (deposits of soluble salts on the surface), cryptoflorescence (accumulation of soluble salts just beneath the surface), surface crust, and interstitial condensation. Moisture penetration into masonry can lead to disintegrating

SEI/ASCE 11-99

TABLE 4.3.1(a). Evaluation of Properties of Structural Steel

References: See numerical references from noted tables in Section 3.7	Table 3.7.6 22	Table 3.7.6 3,10	Table 3.7.6 14,16 17,20	Table 3.7.6 13,15 18,19	Table 3.7.7 1,7,8	Table 3.7.6 1,9, 12	Table 3.7.6 11	Table 3.7.6 4, 5						
Evaluation Procedure / Chemical and Physical Properties	Bend Tests	Chemical Analysis	Fatigue Tests	Fracture Tests	Hardness Tests	Impact Test	Modulus Test	Tension Test						
Carbon Content		●												
Chemical Composition		●												
Ductility	●													
Elongation								●						
Fatigue Properties														
Cycle Counting			●											
Endurance Limit			●											
e-N Curve			●											
Hardness					●									
S-N Curve			●											
Fracture Properties														
Charpy Number						●								
Fracture Toughness				●										
Nil Ductility Trans. Temp.				●										
Modulus of Elasticity							●	●						
Reduction in Area								●						
Tensile Strength								●						
Yield Strength								●						

TABLE 4.3.1(b). Evaluation of Physical Conditions of Structural Steel

References: See Section 3.7 and Table 3.7.8	4,11	7,11	6	3,11 12,13	5,11 12,13	9,14	1,8 11,12 13	14,17 18	2,16 19,20	9,14				
Physical Condition \ Evaluation Procedure	Acoustic Emission	Eddy Current	Hole Drilling	Liquid Penetration	Magnetic Particle	Physical Measurement	Radiography	Structural Analysis	Ultrasonic	Visual Examination				
Bracing of Comp. Elements and Members						•				•				
Cross-Sectional Properties						•								
Deformations						•				•				
Direct Chemical Attack						•								
Electrolytic or Electro-chemical Corrosion						•				•				
Fatigue Cracking	•	•		•	•	•	•		•					
Fire Protection						•				•				
Fracture Cracking	•	•		•	•	•	•		•	•				
Geometry of Structural Components						•				•				
In-Situ Stresses			•											
Laminar Tearing	•	•		•	•	•	•		•	•				
Overall or Local Buckling						•		•		•				
Overstressing								•						

TABLE 4.3.5(a) (1 of 2). Evaluation of Properties of Connectors of Metals

References: See numerical references from noted tables in Section 3.7	Table 3.7.6 3,10	Table 3.7.4 3,23 24	Table 3.7.6 21	Table 3.7.6 1,4,5 11,12 13,15 18,19	Table 3.7.4 6,8 20,22
Evaluation Procedure / Chemical and Physical Properties	Chemical Analysis	Coating Test	Fastener Mechanical Tests	Steel Products Mechanical Tests	Stress Extension Tests
Bolts and Rods (Including Washers and Nuts)					
Chemical Composition	●				
Hardness			●		
Proof Load			●		
Tensile Strength			●		
Rivets					
Chemical Composition	●				
Hardness			●		
Studs					
Chemical Composition	●				
Elongation			●		
Reduction in Area			●		
Tensile Strength			●		
Welds					
Tensile Strength of Filler Material				●	
Yield Strength of Filler Material				●	

TABLE 4.3.5(a) (2 of 2). Evaluation of Properties of Connectors of Metals

References: See numerical references from noted tables in Section 3.7	Table 3.7.6 3,10	Table 3.7.4 3,23 24	Table 3.7.6 21	Table 3.7.6 1,4,5 11,12 13,15 18,19	Table 3.7.4 6, 8 20,22							
Chemical and Physical Properties \ Evaluation Procedure	Chemical Analysis	Coating Test	Fastener Mechanical Tests	Steel Products Mechanical Tests	Stress Extension Tests							
Wire Strand and Wire Rope												
Breaking Strength				•								
Coating Weight		•		•								
Elongation				•								
Stress at 0.7% Extension				•								
Tensile Strength				•	•							

TABLE 4.3.5(b) (1 of 2). Evaluation of Physical Conditions of Connectors of Metals

References: See Section 3.7 and Table 3.7.8	4,11	17,18	7,11	3, 4 12,13	5,11 12,13	9, 14	1, 8 11,12 13	2,16 19,20	9, 14					
Physical Condition \ Evaluation Procedure	Acoustic Emission	Bolt Tightness	Eddy Current	Liquid Penetration	Magnetic Particle	Physical Measurements	Radiographic	Ultrasonic	Visual Examination					
Bolts and Rods (Including Washers and Nuts)														
Condition						•			•					
Corrosion						•			•					
Cross-Sectional Properties						•								
Deformation						•			•					
Dimensions						•		•						
Nonbinding Condition						•								
Notched holes for Movement						•								
Tightness		•												
Rivets														
Condition						•			•					
Corrosion						•			•					
Cross-Sectional Properties						•								
Deformation						•			•					
Dimensions						•								
Soundness									•					
Tightness									•					

TABLE 4.3.5(b) (2 of 2). Evaluation of Physical Conditions of Connectors of Metals

References: See Section 3.7 and Table 3.7.8	4,11	17,18	7,11	3, 4, 12,13	5,11 12,13	9,14	1, 8 11,12 13	2,16 19,20	9,14					
Physical Condition \ Evaluation Procedure	Acoustic Emission	Bolt Tightness	Eddy Current	Liquid Penetration	Magnetic Particle	Physical Measurements	Radiographic	Ultrasonic	Visual Examination					
Studs														
Condition						•			•					
Corrosion						•			•					
Cross-Sectional Properties						•								
Deformation						•			•					
Welds														
Cracks	•		•	•	•		•	•						
Discontinuities	•		•	•	•		•	•						
Length						•								
Location						•			•					
Profile						•			•					
Porosity	•		•	•	•		•	•						
Size						•			•					
Slag Deposits	•		•	•	•		•	•						
Smoothness									•					
Uniformity									•					
Wire Strand and Wire Rope														
Condition						•			•					
Corrosion						•			•					
Cross-Sectional Properties						•								
Length						•								
Tightness									•					

cycles of freezing and thawing and corrosion of masonry ties, anchors, joint reinforcement and reinforcing steel. Freezing and thawing, weathering, and corrosion of embedded metal results in cracking, delamination, erosion, exfoliation and friability of natural stones, pitting of the masonry surface, and spalling.

Structural Incompatibility. Deterioration of masonry can be caused by structural incompatibility including movement of supports, insert slippage, settlement of foundations, overloads and volume changes in the structure due to elastic shortening, creep or thermal changes. Masonry is often used as a finish material in conjunction with structural reinforced concrete or structural steel. The finish masonry is dependent on its attachment and support from the structure. These supports can move (vertical, horizontal, or rotational movement of supporting shelf angles) and inserts supporting shelf angles can slip. The supporting structure can settle (perhaps due to overloads) and undergo volume changes. These movements can result in bowing and bulging, chipping, cracking, detachment, displacement, distortion, sagging, spalling, and warping of the masonry.

Inadequate Architectural Features. Deterioration of masonry can result from inadequate architectural features. These include the condition of the flashing (loss of integrity permitting entrance of moisture and water); inadequate control joints to control and accommodate contraction, settlement, and temperature and moisture volume changes; inadequate expansion joints to control expansion; missing damp proof courses used to prevent capillary flow of moisture; clogged weep holes; and lack of horizontal expansion joints where deflection of the supporting structures would allow mass of masonry above a shelf angle to be transmitted to the masonry below the shelf angle.

Mortar and Grout Deterioration. Masonry units are joined together by mortar. Reinforced masonry is grouted into place (Snell and Rutledge, 1990). Poorly constructed mortar and grout joints can result in misalignment of the masonry units, variation in bed thickness and levelness which can result in cracking of the mortar and grout permitting access for moisture and water. Improperly constructed masonry joints and improper grouting can also permit exposure of masonry ties, anchors, joint reinforcement and reinforcing steel to deterioration.

4.4.2 Properties and Physical Conditions

The ability of masonry to function as a structural material as well as to enclose areas and protect them from the environment indicates that the properties or physical conditions tabulated in Tables 4.4.5(a), 4.4.5(b), 4.4.5(c), 4.4.5(d), 4.4.5(e), 4.4.5(f), and 4.4.5(g) may be considered in evaluating the acceptability of masonry.

4.5 EVALUATION OF STRUCTURAL WOOD

Structural wood includes all wood products used for structural purposes including wood used for connections and mechanical connectors and connections made by gluing. Wood products include solid sawn wood as lumber and timber; structural composite lumber such as structural glued laminated timber and laminated veneer lumber; trusses; and structural wood panel products such as plywood, composite panels and combined wood structural elements.

4.5.1 Wood

Wood includes all wood products and wood used for connections. Wood functions as a structural material in all capacities as beams, columns, diaphragms, bearing plates, or combinations of these to support and distribute vertical and horizontal loads. The ability of wood to function in these capacities is measured by its designated stress grade. The stress grade is a function of the presence or absence of characteristics that may affect the strength. These include knots, checks, shakes, pitchpockets; slope of grain; and number of annual growth rings per inch. The basic allowable stress is a function of the species, the stress grade and the size classification (dimension lumber, post and timbers, etc). Furthermore, the ability of wood to function structurally is related to the moisture content of the wood and its variations, thermal changes either along or cross grain, the direction of application of load to the grain of the wood, existence of preservatives and fire retardants, temperature environment, and the duration of the applied load (intermittent or sustained). Wood is not only susceptible to sudden damage and deterioration but with time it can exhibit permanent deflection, bowing, twisting and shrinkage which all have their effect on the capacity of wood to function as a structural material.

4.5.1.1 Causes of Wood Deterioration
Wood is a biological rather than a manufactured material and accordingly possesses inherent defects as knots, splits and variability in species that must be considered in evaluating engineering properties. Some of these inherent defects can change with time. An-

TABLE 4.4.5(a) (1 of 2). Evaluation of Properties of Masonry Units

References: See numerical references from noted tables in Section 3.7	Table 3.7.9 17,37	Table 3.7.9 19,27	Table 3.7.9 43	Table 3.7.10 12	Table 3.7.9 11,12 36	Table 3.7.3 15,16 17	Table 3.7.9 8, 18 19,28	Table 3.7.10 12
Physical Property / Evaluation Procedure	Compressive Tests	Efflorescence Test	Freezing and Thawing Test	Measurement Tests	Modulus Tests	Moisture-Related Tests	Properties of Finish Tests	Visual Examination
Absorption						•		
Adhesion	•							
Area in Cored Units				•				
Autoclave Crazing							•	
Bulk-Specific Gravity						•		
Chemical Resistance						•		
Compressive Load	•							
Compressive Strength	•							
Dimensions				•				
Distortion				•				
Drying Shrinkage						•		
Efflorescence		•						
Freezing and Thawing			•					
Imperviousness						•		
Initial Rate of Absorption						•		

TABLE 4.4.5(a) (2 of 2). Evaluation of Properties of Masonry Units

References: See numerical references from noted tables in Section 3.7	Table 3.7.9 17,37	Table 3.7.9 19,27	Table 3.7.9 43	Table 3.7.10 12	Table 3.7.9 11,12 36	Table 3.7.3 15,16 17	Table 3.7.9 8, 18 19,28	Table 3.7.10 12
Physical Property \ Evaluation Procedure	Compressive Tests	Efflorescence Test	Freezing and Thawing Test	Measurement Tests	Modulus Tests	Moisture-Related Tests	Properties of Finish Tests	Visual Examination
Modulus of Elasticity					•			
Modulus of Rupture					•			
Moisture Content						•		
Moisture Expansion						•		
Net Area				•				
Opacity							•	
Saturation Coefficient						•		
Secant Modulus of Elasticity						•		
Size				•				
Solubility							•	
Splitting Tensile Strength	•							
Surface Texture							•	
Thermal Expansion							•	
Warpage				•				
Weight Determination							•	
Young Modulus					•			

TABLE 4.4.5(b). Evaluation of Properties of Masonry Assemblages

Physical Property \ Evaluation Procedure	Vertical Load Tests	Transverse Load Tests	Shear Tests	Deformability Tests								
References: See Section 3.7 and Table 3.7.9	31,36 37	30,31 36,41	31,36 41	31,34 36								
Compressive Load	•											
Concentrated Loads	•											
Deformability				•								
Flexural Tensile Strength		•										
Modulus of Rupture		•										
Secant Modulus of Rupture		•										
Shear Strength			•									
Tensile Load		•										
Transverse Load		•										

TABLE 4.4.5(c). Evaluation of Properties of Mortar and Grout

Physical Property \ Evaluation Procedure	Petrographic Tests													
References: See Section 3.7 and Table 3.7.9	5,9,17 29													
Aggregate Characteristics	●													
Mortar and Grout Characteristics	●													
Paste Characteristics	●													
Air Content	●													

TABLE 4.4.5(d) (1 of 2). Evaluation of Physical Conditions of Masonry Units and Masonry Assemblages (Environmental)

References: See numerical references from noted tables in Section 3.7	Table 3.7.10 17,35,37	Table 3.7.3 12,16 27	Table 3.7.3 24,35	Table 3.7.10 3,7,10 17,19, 25,35 37	Table 3.7.10 12	Table 3.7.10 1,17, 18,20 22,26 27,35	Table 3.7.10 4,17 32	Table 3.7.10 3,7,9 17,23 28,35					
Physical Condition \ Evaluation Procedure	Acoustic Impact	Electrical Resistance Probe	Fiber Optics	Gamma Radiography	Physical Measurements	Ultrasonic Pulse Velocity	Visual Examination	Water Penetration					
Air Leakage					•		•						
Blistering							•						
Corrosion		•	•				•						
Coving							•						
Cracking	•	•	•	•	•	•	•						
Crazing							•						
Crumbling							•						
Cryptoflorescence			•				•						
Delamination	•			•		•	•						
Deterioration		•	•				•						
Dew Point					•								
Driving Rain Index					•								
Efflorescence							•						
Erosion							•						
Exfoliation							•						
Flaking							•						
Friability							•						

TABLE 4.4.5(d) (2 of 2). Evaluation of Physical Conditions of Masonry Units and Masonry Assemblages (Environmental)

References: See numerical references from noted tables in Section 3.7 / Evaluation Procedure → Physical Condition ↓	Table 3.7.10 17,37 Acoustic Impact	Table 3.7.3 12,16 27 Electrical Resistance Probe	Table 3.7.3 24 Fiber Optics	Table 3.7.10 3,7,10 17,25 37 Gamma Radiography	Table 3.7.10 12 Physical Measurements	Table 3.7.10 1,17, 18,20 22,26 27 Ultrasonic Pulse Velocity	Table 3.7.10 4,17 32 Visual Examination	Table 3.7.9 3,7,9 17,23 28 Table 3.7.10 28 Water Penetration
Interstitial Condensation		•	•				•	
Lime Run							•	
Peeling							•	
Pitting							•	
Rising Damp		•	•		•		•	
Salt Fretting							•	
Spalling							•	
Staining							•	
Sugaring							•	
Surface Condensation							•	
Surface Crust							•	
Surface Induration							•	
Surface Temperature							•	
Tidemark							•	
Transition Area			•				•	
Warpage							•	
Water Penetration/Permeance		•					•	•
Weathering							•	

TABLE 4.4.5(e). Evaluation of Physical Conditions of Masonry Units and Masonry Assemblages (Structural)

References: See Section 3.7 and Table 3.7.10 / Physical Condition \ Evaluation Procedure	Physical Measurements	Visual Examination												
	12	12												
Bowing	•	•												
Bulging	•	•												
Chipping		•												
Condition of Lintels (Shelf Angles)		•												
Cracking	•	•												
Deflection/Settlement	•	•												
Detachment		•												
Displacement	•	•												
Distortion	•	•												
Plumbness	•	•												
Sagging	•	•												
Spalling		•												
Ties and Anchors		•												
Warping	•	•												

TABLE 4.4.5(f). Evaluation of Physical Conditions of Masonry Units and Masonry Assemblages (Architectural)

References: See Section 3.7 and Table 3.7.10 / Physical Condition \ Evaluation Procedure	Physical Measurements	Visual Examination												
	12	12												
Condition of Flashing	●	●												
Condition of Roofing		●												
Control Joints	●	●												
Damp Course		●												
Dampproofing		●												
Expansion Joints		●												
Gutters and Downspouts		●												
Location		●												
Horizontal Expansion Joint		●												

SEI/ASCE 11-99

123

TABLE 4.4.5(g). Evaluation of Physical Conditions of Mortar and Grout

References: See numerical references from noted tables in Section 3.7	Table 3.7.3 8, 9, 27	Table 3.7.10 24, 35	Table 3.7.10 3, 7, 10, 17, 25, 35, 37	Table 3.7.3 12	Table 3.7.10 12								
Physical Condition \ Evaluation Procedure	Cover Meter	Fiber Optics	Gamma Radiography	Physical Measurement	Visual Examination								
Alignment				•	•								
Bed Joint Levelness				•	•								
Bed Width Variation				•	•								
Condition					•								
Cracking		•	•	•	•								
Exposure of Connectors and Reinforcement	•	•	•										
Friability				•									
Hardness				•									
Pitting				•									
Sandiness				•									
Voids				•									

other biological characteristic is that wood is susceptible to attack by various insects and to decay. Deterioration of wood in buildings is affected by the duration of load and load history, moisture content, damage from accidents, fire, insect attack, and decay.

Accident Damage. Members near traffic areas in garages and loading docks, are very susceptible to damage resulting from impact. Damage may range from split-off corners or fractures to total demolition of individual members. Similar damage may have also occurred during shipment and improper handling of the members during erection and the damage never corrected.

Decay. One of the main causes of wood deterioration is decay (Emmerich, 1986). Decay is caused by fungi which can manifest itself as molds, stains and decay fungi. Decay fungi is the primary damaging fungi. They consume the wood as food and thus reduces its strength. For decay fungi to exist, the wood must be available as food for the fungi, the moisture content of the wood must be above 20%, temperatures must be approximately 70°F to 90°F (21°C to 32°C), and air is required for fungi growth. There are three stages of decay development: (1) incipient decay where the infected wood is subjected to decay fungi in long finger-like growths. Incipient decay is difficult to identify, (2) intermediate decay where the wood becomes soft and punky with obvious loss of strength, and (3) advanced decay where no strength remains and the wood is literally dissolved away. Wood destroying fungi are reproduced by spores. These are wind borne and are carried to where conditions are right, where the spores generate and affect new wood.

Duration of Load. Load duration is the length of time a member supports a load. The load required to produce failure over a long period of time is less than the load required to produce failure over a shorter period of time. Present-day design specifications recognize this condition and allowable stresses are adjusted for duration of load. For some of the older wood structures, this adjustment was not considered in the design and the structures became overloaded with time. However, this overload is in the magnitude of approximately 10%—not a serious problem except when other factors affect the allowable stresses and the cumulative effect must be evaluated.

Fire. A characteristic of wood is that it develops a char, which may affect the rate of fire damage based on moisture content and density of wood. Based on the characteristic of a fire, damage may vary across a member and therefore requires careful investigation. Damage to the wood may be extended beyond the obvious area of charring.

Insect Attack. Insect damage can occur in standing trees, logs, and in unseasoned and seasoned wood. Insect holes are generally classified as pin holes, grub holes and powderpost holes (created by powderpost larvae). When systems of insect tunnels and chambers become interconnected, they are termed a network of galleries. Most of the damage done to wood structures are from subterranean termites. Termites are more prevalent in warmer climates and infestation is greatest in poorly drained areas and wood close to the ground since subterranean termites develop and maintain their colonies in the ground. Damage by marine borers to wood foundation piles in salt or brackish waters is practically worldwide. The rapidity of attack depends upon local conditions and the kinds of borers present. Untreated wood piles can be destroyed completely in less than a year in the warmer climates. Loss of protective coatings with time can make piles and wood in water susceptible to marine borer attack. Interestingly, pollution of the waterways has provided some protection against marine borers and as harbors and rivers are cleaned up, marine borers have moved back into these areas with a vengeance.

Moisture Content. Moisture has a dramatic effect on the properties of wood. If wood changes moisture content after installation, the resulting change in dimension can cause distortion and twisting. The change in dimension only occurs at moisture contents below the fiber saturation point (approximately 30% moisture content). A decrease below this point will increase the strength of wood but it will also cause it to shrink resulting in checks and splits caused by differential shrinkage. Grading of wood and establishment of allowable stresses takes into consideration the number of checks and splits. Accordingly, as these increase, the allowable stresses decrease even though the actual wood material between the checks and splits may have a higher strength. A reduction in allowable stresses is equivalent to overloading the structure. A change in moisture content above the fiber saturation point does not affect strength or shrinkage. However, increases below this point result in lower moduli of elasticities and lower permitted allowable stresses. Thus an increase in moisture content from the design and installed value up to the fiber saturation point can result in equivalent overloading of the structure.

4.5.1.2 Properties and Physical Conditions

As a structural material the properties and physical conditions of wood that are summarized in Tables 4.5.1(a) and 4.5.1(b) should be considered in evaluating the acceptability of existing wood members and connections.

4.5.2 Connections

Connections include the component metal or wood shapes used to attach one structural element to another. All applicable provisions of Sections 4.3.1 and 4.5.1 apply as well as the applicable portions of Tables 4.5.3(a) and 4.5.3(b).

4.5.3 Connectors

Connectors are the means of making connections. Connectors are used to fasten wood members together. They include bolts and rods, timber connectors, nails, and screws. They are required to transmit forces between the elements that are being connected. This requirement indicates that the properties or physical conditions tabulated in Tables 4.5.3(a) and 4.5.3(b) may be considered in evaluating the acceptability of existing connectors.

4.6 OTHER MATERIALS

Materials other than those specifically referenced in this standard, or combinations of materials that have been referenced, may be evaluated by a qualified professional engineer using the same general principles as established herein on a case-by-case basis.

4.7 EVALUATION OF COMPONENTS AND SYSTEMS

4.7.1 Basis of Evaluation

The evaluation should be consistent with the stated purpose or purposes of the assessment (see Section 1.2).

4.7.2 Considerations

4.7.2.1 Interpretation of Data

Data should be interpreted statistically when appropriate and feasible.

4.7.2.2 Material Properties

Physical, chemical, and other properties of materials and components may be determined by laboratory testing.

4.7.2.3 Sample Significance

Due consideration and weight should be given to the importance of an element or unit being tested, evaluated, etc., with respect to the other elements or units in the overall system being evaluated. It is both permissible and desirable to weigh this significance mathematically. A minimum number of samples must be taken to ensure statistical significance and to establish confidence level. In utilizing limit state concepts, tests on a significant number of actual material samples to determine compressive and tensile strengths could justify a revision in the phi reduction factor.

4.7.2.4 Measure of Safety

It is permissible and desirable to use limit state concepts for determination of reliability (or safety) index in lieu of a conventional factor of safety. An increase or decrease in the normal factor of safety may be considered. The evaluation should determine what is an appropriate factor of safety depending upon the existing loads, the usage, and the type of analysis and investigation conducted.

4.7.2.5 System Approach

Approaches that give cognizance to system evaluation rather than to component or subsystem evaluation should be utilized. The use of load redistribution is acceptable as long as there is a physical redistribution system available.

4.7.2.6 Load Tests

A load test provides only a limited amount of information on the performance under a specific load. There are many other factors that must be considered and evaluated to determine the acceptability of a structure. Load tests should be conducted with analytical modeling of the test area and respective verification of the test results. The performance and interpretation of loading tests require special knowledge.

4.7.2.7 Stiffness

Stiffness is the resistance to deformation of a member or structure measured by the ratio of the applied force to the corresponding displacement. A change in stiffness of an existing building or a change in the distribution of the stiffening elements,

TABLE 4.5.1(a). Evaluation of Properties of Wood

References: See numerical references from noted tables in Section 3.7	Table 3.7.11 1,2 10,11 12	Table 3.7.11 1,2 10,11 12	Table 3.7.11 1,5 8,13	Table 3.7.11 1,2 10,11 12	Table 3.7.13 ALL REF.	Table 3.7.11 1,5 13	Table 3.7.11 1,2 10,11	Table 3.7.12 1,6 20,26	Table 3.7.12 14,26 45	Table 3.7.11 1,3 10,11 12	Table 3.7.12 5,9 16,17 33,34	Table 3.7.11 1,2 10,11 12	Table 3.7.12 1,28 45,46	Table 3.7.11 3,11 12,15	Table 3.7.11 3,11 12,15
Mechanical and Physical Property \ **Evaluation Procedure**	Bending Tests	Compression Tests	Density Tests	Fatigue Tests	Load Tests	Moisture Tests	Modulus Tests	Penetration Test	Radiographic	Shear Tests	Stress Wave Propagation	Tension Tests	Visual Examination	Visual Grading	Species Identification
Bending Strength	●				●										
Classification													●	●	
Compression Parallel to Grain		●						●			●				
Compression Perpendicular to Grain		●													
Density			●					●	●						
Tensile Strength				●											
Fiber Saturation Point						●									
Grade													●	●	
Growth Characteristics															
Heartwood													●		
Knots									●				●	●	
Sapwood													●		
Slope of Grain		●							●				●	●	
Modulus of Elasticity	●						●				●				
Shear Strength										●					
Species													●		●
Tensile Strength Parallel to Grain												●			
Tensile Strength Perpendicular to Grain												●			

GUIDELINE FOR STRUCTURAL CONDITION ASSESSMENT OF EXISTING BUILDINGS

TABLE 4.5.1(b) (1 of 2). Evaluation of Physical Conditions of Wood

Physical Condition \ Evaluation Procedure	Coring	Decay Tests	Drilling	Glue Tests	Moisture Meter	Penetration Test	Physical Measurement	Probing	Radiography	Sounding	Species Identification	Stress Wave Propagation	Termite Tests	Visual Examination	Wood Treatment Tests	Visual Grading
References: See numerical references from noted tables in Section 3.7	Table 3.7.12 6, 8, 13, 20, 32	Table 3.7.12 1, 6, 13,20, 26,33, 34	Table 3.7.12 20,32	Table 3.7.12 1	Table 3.7.12 1, 2,19, 20,31, 35	Table 3.7.12 1, 6, 20,26	Table 3.7.11 1,2,3, 15,26	Table 3.7.12 6, 20, 26	Table 3.7.12 14,26, 45	Table 3.7.12 1,20	Table 3.7.11 3, 11, 12	Table 3.7.12 5, 9, 16,17, 33,34	Table 3.7.12 1	Table 3.7.12 1, 28, 45,46	Table 3.7.11 5, 7, 9	Table 3.7.11 3, 11, 12,14, 15
Adhesive																
Chemical Exposure											•			•		
Chemical Treatments															•	
Condition of Glue				•												
Cross-Sectional Properties							•									
Decay and Decay Exposure		•				•		•	•		•	•		•		
Deflection							•							•		
Fire Retardants and Preservatives															•	
Fungal Damage								•			•			•		
Insect Infestation and Animal Attack								•	•		•		•	•		
Moisture and its Effect																
Pattern of Exposure					•									•		
Effects of Changing Moisture Content																
Checks and End Splits									•					•		•
Shrinkage														•		
Twisting							•							•		
Warping (Bowing, Crooking, and Cupping)							•							•		•
Moisture Content					•			•								

SEI/ASCE 11-99

TABLE 4.5.1(b) (2 of 2). Evaluation of Physical Conditions of Wood

References: See numerical references from noted tables in Section 3.7	Table 3.7.12 6, 8, 13, 20 32	Table 3.7.12 1, 6 13,20 26,33 34	Table 3.7.12 20,32	Table 3.7.12 1	Table 3.7.12 1, 2,19 20,31 35	Table 3.7.11 1, 6 20,26	Table 3.7.12 1,2,3 15,26	Table 3.7.12 6, 20 26	Table 3.7.12 14,26 45	Table 3.7.12 1,20	Table 3.7.11 3, 11 12	Table 3.7.12 5, 9 16,17 33,34	Table 3.7.12 1	Table 3.7.12 1, 28 45,46	Table 3.7.11 5, 7, 9	Table 3.7.11 3, 11 12,14 15
Physical Condition \ Evaluation Procedure	Coring	Decay Tests	Drilling	Glue Tests	Moisture Meter	Penetration Test	Physical Measurement	Probing	Radiography	Sounding	Species Identification	Stress Wave Propagation	Termite Tests	Visual Examination	Wood Treatment Tests	Visual Grading
Physical Damage	●		●				●	●	●					●		
Rot		●				●		●	●				●	●		
Seasoning					●									●		
Temperature Environment														●		
Time Load Duration (time creep)							●							●		
Weathering														●		

TABLE 4.5.3(a) (1 of 2). Evaluation of Properties of Connectors of Wood

Chemical and Physical Properties \ Evaluation Procedure	Chemical Analysis	Delamination Tests	Fastener Mechanical Tests	Hanger Tests	Hardness Tests	Lateral Resistance Tests	Tension Tests	Timber Joint Tests	Withdrawal Tests
References: See numerical references from noted tables in Section 3.7	Table 3.7.6 3,10	Table 3.7.12 12,44	Table 3.7.11 4, 14	Table 3.7.11 4	Table 3.7.6 21	Table 3.7.11 4, 15	Table 3.7.6 21	Table 3.7.11 4	Table 3.7.11 4,15
Bolts and Rods (Including Lag Bolts, Washers, and Nuts)									
Chemical Composition	●								
Hardness			●		●				
Proof Load			●						
Tensile Strength			●						
Joist and Purlin Hangers									
Torsional Moment Capacity				●					
Vertical Load Capacity				●					
Resistance to Delamination		●							
Timber Connectors (Splits Rings, Toothed Rings, Shear Plates)									
Chemical Composition	●								
Elongation							●		
Tensile Strength							●		
Timber Joint Compressive Strength								●	
Timber Joint Tensile Strength								●	
Yield Strength							●		

TABLE 4.5.3(a) (2 of 2). Evaluation of Properties of Connectors of Wood

References: See numerical references from noted tables in Section 3.7 / Evaluation Procedure vs. Chemical and Physical Properties	Table 3.7.6 3, 10 Chemical Analysis	Table 3.7.12 12, 44 Delamination Tests	Table 3.7.11 4, 14 Fastener Mechanical Tests	Table 3.7.11 4 Hanger Tests	Table 3.7.6 21 Hardness Tests	Table 3.7.11 4, 15 Lateral Resistance Tests	Table 3.7.6 21 Tension Tests	Table 3.7.11 4 Timber Joint Tests	Table 3.7.11 4, 15 Withdrawal Tests					
Wood Nails														
Chemical Composition	•													
Lateral Resistance						•								
Withdrawal Strength								•	•					
Wood Screws (Including Lag Screws and Spiral Dowels)														
Chemical Composition	•													
Lateral Resistance						•								
Withdrawal Strength								•	•					

TABLE 4.5.3(b). Evaluation of Physical Conditions of Connectors of Wood

Physical Condition \ Evaluation Procedure	Physical Measurements	Visual Examination	Corrosion Rate											
References: See Section 3.7 and Table 3.7.8	15	15	15											
Angle of Axis of Connector to Grain	●													
Condition of Connector		●												
Condition of Glue		●												
Corrosion		●	●											
Cross-Section Properties	●													
Deformation	●	●												
Dimensions	●													
Eccentricity	●	●												
Elongated Bolt Holes	●	●												
Nonbinding Condition		●												
Pentration of Nails and Screws	●													
Splits At Connection		●												
Tightness		●												

as well as in the stiffness of these individual elements, can materially affect both the distribution of the applied load and the resulting deformations. Overall structural stiffness is made up of many components, including the structural frame, floor systems, exterior walls and panels, architectural in-fill walls, stairs, and composite action of structural and nonstructural elements. Damage or alteration to the components of stiffness, which could be construed to be only architectural in nature, might in fact seriously affect the load-carrying capacity of the structure or of its component elements.

4.7.2.8 Stability

Stability is a measure of the capacity of a structure to resist additional load. Loss of stability is indicated by exhaustion of this resistance. The factor-of-safety against collapse is the ratio of the maximum load a structure can support to the estimated loads. Stability is closely related to stiffness in that many of the components of building stiffness contribute to the overall building stability or factor-of-safety against collapse. Accordingly those same structural elements that affect stiffness will also affect stability and should be evaluated as part of a structural condition assessment to determine the effects on the building's stability.

4.7.2.9 Overturning and Sliding

Overturning and sliding of structures under the action of lateral loads shall be computed and evaluated against appropriate factors of safety.

4.8 INTERPRETATION

Interpretation of the results of the evaluation is based on the professional experience and judgmental opinion of individual experts who are each responsible for the evaluation. Because these experts will be accepting the professional responsibility for the overall evaluation, interpretation is not standardized herein.

4.9 REFERENCES FOR EVALUATION PROCEDURES

4.9.1 References

ASTM Committee on Terminology. (1990). "Compilation of ASTM Standard Definitions," PCN 03-001090-42, American Society for Testing and Materials, West Conshohocken, Pennsylvania.

ASTM A143. (1994). "Standard Practice for Safeguarding Against Embrittlement of Hot-Dip Galvanized Structural Steel Products and Procedures for Detecting Embrittlement," 1998 Annual Book of ASTM Standards, *Vol. 1.06*, American Society for Testing and Materials, West Conshohocken, Pennsylvania, 23–25.

Bennett, E. W. (1982). "Fatigue Tests of Spliced Reinforcement in Concrete Beams," Fatigue of Concrete Structures, edited by S.P. Shah, *Publication SP-75*, Paper SP 75-8, American Concrete Institute, Farmington Hills, Michigan, 177–193.

Borgard, B., Warren, C., Somayaji, S., and Heidersbach, R. (1990). "Mechanisms of Corrosion of Steel in Concrete," *Symposium on Corrosion Rates of Steel in Concrete, ASTM STP 1065*, edited by N. S. Berke, V. Chaker, and D. Whiting, American Society for Testing and Materials, West Conshohocken, Pennsylvania, 174–188.

Brown, B. F. (1972). "A Preface to the Problem of Stress Corrosion Cracking," Stress Corrosion Cracking of Metals—A State of the Art, *ASTM STP 518*, American Society for Testing and Materials, West Conshohocken, Pennsylvania, 3–15.

Cornet, J., Pirtz, D., Polivka, M., Gau, Y., and Shimizu, A. (1980). "Laboratory Testing and Monitoring of Stray Current Corrosion of Prestressed Concrete in Seawater," *Symposium on Corrosion of Reinforcing Steel in Concrete, ASTM STP 713*, edited by D. E. Tonini, and J. M. Gaidis, American Society for Testing and Materials, West Conshohocken, Pennsylvania, 17–31.

Emmerich, R. W. (1986). "Decay in Wood Structures," Evaluation and Upgrading of Wood Structures, Case Studies, *Proc. of a session at Structures Congress '86 sponsored by the Structural Division*, edited by V. K. A. Gopu, New Orleans, Louisiana, September 15–18, American Society of Civil Engineers, Reston, Virginia, 28–34.

Fontana, M. G., and Greene, N. D. (1978). "Corrosion Engineering," McGraw-Hill, Inc., New York, New York.

Gerwick, B. C., Jr. (1975). "Practical Methods of Ensuring Durability of Prestressed Concrete Ocean Structures," Durability of Concrete, *ACI Publication SP-47*, American Concrete Institute, Farmington Hills, Michigan, 317–324.

Griess, J. C., and Naus, D. J. (1980). "Corrosion of Steel Tendons Used in Prestressed Concrete Pressure Vessels," *Symposium on Corrosion of Reinforc-*

ing Steel in Concrete, *ASTM STP 713*, edited by D. E. Tonini and J. M. Gaidis, American Society for Testing and Materials, West Conshohocken, Pennsylvania, 32–50.

Grimm, C. T. (1985b). "Durability of Brick Masonry: A Review of the Literature," Masonry, Research, Application, and Problems, *ASTM STP 871*, edited by J. T. Conway and J. C. Grogan, American Society for Testing and Materials, West Conshohocken, Pennsylvania, 202–238.

Hansen, W. C. (1966). "Concrete Aggregates—Chemical Reactions," Significance of Tests and Properties of Concrete and Concrete Making Materials, *ASTM STP 169A*, American Society for Testing and Materials, West Conshohocken, Pennsylvania, 487–496.

Hawkins, N. M., and Shah, S. P. (1982). "American Concrete Institute Considerations for Fatigue," *IABSE Colloquium Lausanne, IABSE Reports*, Volume 37, International Association for Bridge and Structural Engineering, Zurich, Switzerland, 41–50.

Helgason, T., Hanson, J. M., Somes, N. F., Corley, W. G., and Hohnestad, E. (1976). "Fatigue Strength of High-Yield Reinforcing Bars," NCHRP Bulletin 164, Transportation Research Board, National Research Council, Washington, D.C.

Hobbs, D. W. (1988). "Alkali-Silica Reaction in Concrete," Thomas Telford, Ltd., London.

Isecke, B. (1982). "Collapse of the Berlin Congress Hall Prestressed Concrete Roof," *Materials Performance*, National Association of Corrosion Engineers, Houston, Texas, December, 36–39.

Kordina, K., and Quast, U. (1982). "Fatigue Tests on Couplings of Tendons under Service Conditions," *IABSE Symposium, Washington, D.C., Maintenance, Repair and Rehabilitation of Bridges, Final Report, IABSE Reports*, Vol. 39, International Association for Bridge and Structural Engineering, Zurich, Switzerland, 155–160.

Lenschow, R. (1982). "Fatigue of Concrete Structures," *IABSE Colloquium Lausanne, IABSE Reports*, Volume 37, International Association for Bridge and Structural Engineering, Zurich, Switzerland, 15–24.

Locke, C. E. (1986). "Corrosion of Steel in Portland Cement Concrete: Fundamental Studies," Corrosion Effect of Stray Currents and the Techniques for Evaluating Corrosion of Rebars in Concrete, *ASTM STP 906*, edited by V. Chaker, American Society for Testing and Materials, West Conshohocken, Pennsylvania, 5–13.

Mattsson, E. (1982). "The Atmospheric Corrosion Properties of Some Common Structural Metals—A Comparative Study," *Materials Performance*, National Association of Corrosion Engineers, Houston, Texas, July, 9–19.

McCoy, W. J. (1966). "Other Materials—Mixing and Curing Water for Concrete," Significance of Tests and Properties of Concrete and Concrete Making Materials, *ASTM STP 169A*, American Society for Testing and Materials, West Conshohocken, Pennsylvania, 515–521.

Mehta, P. K. (1977). "Effect of Cement Composition on Corrosion of Reinforcing Steel in Concrete," Chloride Corrosion of Steel in Concrete, *ASTM STP 629*, edited by D. E. Tonini and S. W. Dean, American Society for Testing and Materials, West Conshohocken, Pennsylvania, 12–19.

Mehta, P. K., and Polivka, M. (1975). "Sulfate Resistance of Expansive Cement Concretes," Durability of Concrete, *ACI Publication SP-47*, American Concrete Institute, Farmington Hills, Michigan, 367–379.

Munse, W. H. (1964). "Fatigue of Welded Steel Structures," edited by L. M. Grover, Welding Research Council, New York, New York.

Naaman, A. E. (1982). "Fatigue in Partially Prestressed Concrete Beams," Fatigue of Concrete Structures, edited by S. H. Shah, *Publication SP-75*, Paper SP75-2, American Concrete Institute, Farmington Hills, Michigan, 25–46.

Naaman, A. E. (1989). "Fatigue of Reinforcement in Partially Prestressed Beams," Structural Materials, *Proc. of the sessions related to structural materials at Structures Congress '89*, edited by J. F. Orofino, ASCE, San Francisco, California, May 1–5, 377–381.

Nurnberger, V. (1982). "Fatigue Resistance of Reinforcing Steel," *IABSE Colloquium, Lausanne, Fatigue of Steel and Concrete Structures, IABSE Reports*, Volume 37, International Association for Bridge and Structural Engineering, Zurich, Switzerland, 213–220 (in German).

Peterson, P. C. (1977). "Concrete Bridge Deck Deterioration in Pennsylvania," Chloride Corrosion of Steel in Concrete, *ASTM STP 629*, edited by D. E. Tonini and S. W. Dean, American Society for Testing and Materials, West Conshohocken, Pennsylvania, 61–68.

Raths, C. H. (1985). "Brick Masonry Wall Nonperformance Causes," Masonry, Research, Application, and Problems, *ASTM STP 871*, edited by J. T. Conway and J. C. Grogan, American Society for Testing and Materials, West Conshohocken, Pennsylvania, 182–201.

Raymond, L. (1988). "Hydrogen Embrittlement Test Methods: Current Status and Projections," Hydrogen Embrittlement: Prevention and Control, *ASTM STP 962*, edited by L. Raymond, American Society for Testing and Materials, West Conshohocken, Pennsylvania, 10–16.

Reading T. J. (1975). "Combating Sulfate Attack in Corps of Engineers Concrete Construction," Durability of Concrete, *ACI Publication SP-47*, American Concrete Institute, Farmington Hills, Michigan, 343–366.

Rothman, P. S., and Price, R. E. (1986). "Detection and Considerations of Corrosion Problems of Prestressed Concrete Cylinder Pipe," Corrosion Effect of Stray Currents and the Techniques for Evaluating Corrosion of Rebars in Concrete, *ASTM STP 906*, edited by V. Chaker, American Society for Testing and Materials, West Conshohocken, Pennsylvania, 92–107.

Scully, J. R., and Moran, P. J. (1988). "The Hydrogen Embrittlement Susceptibility of Ferrous Alloys: The Influence of Strain on Hydrogen Entry and Transport," Hydrogen Embrittlement: Prevention and Control, *ASTM STP 962*, edited by L. Raymond, American Society for Testing and Materials, West Conshohocken, Pennsylvania, 387–402.

Sheinker, A. A., and Wood, J. D. (1972). "Stress Corrosion Cracking of a High Strength Steel," Stress Corrosion Cracking of Metals—A State of the Art, *ASTM STP 518*, American Society for Testing and Materials, West Conshohocken, Pennsylvania, 16–38.

Snell, L. M., and Rutledge, R. B. (1990). "Procedures to Locate Steel in Concrete Masonry Structures," *Proc., Non-Destructive Evaluation of Civil Structures and Materials*, edited by B. A. Suprenant, S. Sture, J. L. Noland, and M. P. Schuller, University of Colorado, Boulder, Colorado, October, 351–360.

Struble, L., and Diamond, S. (1986). "Influence of Cement Alkali Distribution on Expansion Due to Alkali-Silica Reaction," Alkalies in Concrete, *ASTM STP 930*, edited by V. H. Dodson, American Society for Testing and Materials, West Conshohocken, Pennsylvania, 31–45.

Swamy, R. N., and Al-Asali, M. M. (1986). "Influence of Alkali-Silica Reaction on the Engineering Properties of Concrete," Alkalies in Concrete, *ASTM STP 930*, edited by V. H. Dodson, American Society for Testing and Materials, West Conshohocken, Pennsylvania, 69–86.

Tassi, G., and Magyari, B. (1982). "Fatigue of Reinforcements with Pressed Sleeve Splices," *IABSE Colloquium, Lausanne, Fatigue of Steel and Concrete Structures, IABSE Reports*, Vol. 37, International Association for Bridge and Structural Engineering, Zurich, Switzerland, 265–271.

Tremper, B. (1966). "Hardened Concrete—Corrosion of Reinforcing Steel," Significance of Tests and Properties of Concrete and Concrete Making Materials, *ASTM STP 169A*, American Society for Testing and Materials, West Conshohocken, Pennsylvania, 220–229.

U.S. Department of Interior. (1981). "Concrete Manual," 8th Edition, U.S. Department of Interior, Water and Power Resources Services.

Verbeck, G. J. (1975). "Mechanisms of Corrosion of Steel in Concrete," *ACI SP-49*, Corrosion of Metals in Concrete, edited by L. Pepper, R. G. Pike, and J. A. Willett, American Concrete Institute, Farmington Hills, Michigan, 21–38.

Walls, J. C., Sanders, W. W., Jr., and Munse, W. H. (1965). "Fatigue Behavior of Butt-Welded Reinforcing Bars in Reinforced Concrete Beams," ACI Journal, *Proc.*, Vol. 62, No. 2, Title No. 62-10, American Concrete Institute, Farmington Hills, Michigan, February, 169–192.

4.9.2 Supplemental References

Andersen, K. T., and Thaulow, N. (1990). "The Study of Alkali-Silica Reactions in Concrete by the Use of Fluorescent Thin-Sections," Petrography Applied to Concrete and Concrete Aggregates, *ASTM STP 1061*, edited by B. Erlin and D. Stark, American Society for Testing and Materials, West Conshohocken, Pennsylvania, 71–89.

Baker, E. A., Money, K. L., and Sanborn, C. B. (1977). "Marine Corrosion Behavior of Bare and Metallic-Coated Steel Reinforcing Rods in Concrete," Chloride Corrosion of Steel in Concrete, *ASTM STP 629*, edited by D. E. Tonini and S. W. Dean, American Society for Testing and Materials, West Conshohocken, Pennsylvania, 30–50.

Berke, N. S., Shen, D. F., and Sundberg, K. M. (1990). "Comparison of The Polarization Resistance Technique to the Macrocell Corrosion Technique," *Symposium on Corrosion Rates of Steel in Concrete, ASTM STP 1065*, edited by N. S. Berke, V. Chaker, and D. Whiting, American Society for Testing and Materials, West Conshohocken, Pennsylvania, 38–51.

Bloem, D. L. (1966). "Concrete Aggregates—Soundness and Deleterious Substances," Significance of Tests and Properties of Concrete and Concrete Making Materials, *ASTM STP 169A*, American Society for Testing and Materials, West Conshohocken, Pennsylvania, 497–512.

Booth, G. S., and Wylde, J. G. (1990). "Procedural Considerations Relating to the Fatigue Testing of Steel Weldments," Fatigue and Fracture Testing of Weldments, *ASTM STP 1058*, edited by J. M. Potter and H. I. McHenry, American Society for Testing and Materials, West Conshohocken, Pennsylvania, 3–15.

Burget, W., and Blauel, J. G. (1990). "Fracture Toughness of Manual Metal-Arc and Submerged Arc Welded Joints in Normalized Carbon-Manganese Steels," Fatigue and Fracture Testing of Weldments, *ASTM STP 1058*, edited by J. M. Potter and H. I. McHenry, American Society for Testing and Materials, West Conshohocken, Pennsylvania, 272–299.

Cady, P. D. (1977). "Corrosion of Reinforcing Steel in Concrete—A General Overview of the Problem," Chloride Corrosion of Steel in Concrete, *ASTM STP 629*, edited by D. E. Tonini and S. W. Dean, American Society for Testing and Materials, West Conshohocken, Pennsylvania, 3–11.

Clark, W. G., Jr., and Landes, J. D. (1976). "An Evaluation of Rising Load K_{iscc} Testing," Stress Corrosion—New Approaches, *ASTM STP 610*, edited by H. L. Craig, Jr., American Society for Testing and Materials, West Conshohocken, Pennsylvania, 108–127.

Cook, H. K., and McCoy, W. J. (1977). "Influence of Chloride in Reinforced Concrete," Chloride Corrosion of Steel in Concrete, *ASTM STP 629*, edited by D. E. Tonini and S. W. Dean, American Society for Testing and Materials, West Conshohocken, Pennsylvania, 20–29.

Corley, W. G., Hanson, J. M., and Helgason, T. (1982). "Background of American Design Procedure for Fatigue of Concrete," *Proc., IABSE Colloquium of Steel and Concrete Structures*, Lausanne, Switzerland.

Cornet, I., and Bresler, B. (1980). "Critique of Testing Procedures Related to Measuring the Performance of Galvanized Steel Reinforcement in Concrete," *Symposium on Corrosion of Reinforcing Steel in Concrete, ASTM STP 713*, edited by D. E. Tonini and J. M. Gaidis, American Society for Testing and Materials, West Conshohocken, Pennsylvania, 160–195.

Cramer, S. M., and Goodman, J. R. (1985). "Predicting Tensile Strength of Lumber," *Proc., Fifth Nondestructive Testing of Wood Symposium*, Pullman, Washington, September 9–11, Conferences and Institutes, Washington State University, Pullman, Washington, 525–545.

Dean, S. W., Jr. (1976). "Review of Recent Studies on the Mechanism of Stress Corrosion Cracking in Austenitic Stainless Steels," Stress Corrosion—New Approaches, *ASTM STP 610*, edited by H. L. Craig, Jr., American Society for Testing and Materials, West Conshohocken, Pennsylvania, 308–337.

Dillard, J., Glanville, J., Osiroff, T., and Weyers, R. (1991). "Surface Characterization of Reinforcing Steel and the Interaction of Steel With Inhibitors in Pore Solution," *Presented at the 70th Annual Meeting, Paper No. 910272*, Transportation Research Board, Washington, D.C., January.

Dressman, S., Osiroff, T., Dillard, J., Glanville, J., and Weyers, R. (1991). "A Screening Test For Rebar Corrosion Inhibitors," *Presented at the 70th Annual Meeting, Paper No. 910281*, Transportation Research Board, Washington, D.C., January.

Fairchild, D. P. (1990). "Fracture Toughness Testing of Weld Heat-Affected Zones in Structural Steel," Fatigue and Fracture Testing of Weldments, *ASTM STP 1058*, edited by J. M. Potter and H. McHenry, American Society for Testing and Materials, West Conshohocken, Pennsylvania, 117–141.

Graham, J. R., and Backstrom, J. E. (1975). "Influence of Hot Saline and Distilled Waters on Concrete," Durability of Concrete, *ACI Publication SP-47*, American Concrete Institute, Farmington Hills, Michigan, 325–341.

Grimm, C. T. (1985). "Corrosion of Steel in Brick Masonry," Masonry, Research, Application, and Problems, *ASTM STP 871*, edited by J. T. Conway and J. C. Grogan, American Society for Testing and Materials, West Conshohocken, Pennsylvania, 67–87.

Grimm, C. T. (1988). "Statistical Primer for Brick Masonry," Masonry: Materials, Design, Construction, & Maintenance, *ASTM STP 992*, edited by H. A. Harris, American Society for Testing and Materials, West Conshohocken, Pennsylvania, 169–192.

Hansson, C. M., and Sørensen, B. (1990). "The Threshold Concentration of Chloride in Concrete for the Initiation of Reinforcement Corrosion," *Symposium on Corrosion Rates of Steel in Concrete, ASTM STP 1065*, edited by N. S. Berke, V. Chaker, and D. Whiting, American Society for Testing and Materials, West Conshohocken, Pennsylvania, 3–16.

Hartt, W. H. (1990). "Corrosion Fatigue Testing of Steels as Applicable to Offshore Structures," Corrosion in Natural Waters, *ASTM STP 1086*, edited by C. H. Baloun, American Society for Testing and Materials, West Conshohocken, Pennsylvania, 54–69.

Idorn, G. M., and Roy, D. M. (1986). "Opportunities with Alkalies in Concrete Testing, Research, and Engineering Practice," Alkalies in Concrete,

ASTM STP 930, edited by V. H. Dodson, American Society for Testing and Materials, West Conshohocken, Pennsylvania, 5–15.

Link, L. R. (1990). "Fatigue Crack Growth of Weldments," Fatigue and Fracture Testing of Weldments, *ASTM STP 1058*, edited by J. M. Potter and H. I. McHenry, American Society for Testing and Materials, West Conshohocken, Pennsylvania, 16–33.

Morf, U. (1989). "Durability of High Strength Bars and Wires in Tension," *IABSE Symposium, Lisbon, Durability of Structures, IABSE Reports*, Vol. 57/1, International Association for Bridge and Structural Engineering, Zurich, Switzerland, September 6–8, 199–204 (in German).

Powers, T. C. (1975). "Freezing Effects In Concrete," Durability of Concrete, *ACI Publication SP-47*, American Concrete Institute, Farmington Hills, Michigan, 1–11.

Rabbers, F., and Norman L. (1986). "Alterations to Old Wood Building," *Evaluation and Upgrading of Wood Structures, Case Studies, Proc. of a session at Structures Congress '86 sponsored by the Structural Division*, edited by V. K. A. Gopu, New Orleans, Louisiana, September 15–18, American Society of Civil Engineers, Reston, Virginia, 96–107.

Robinson, G. C., and Brown, R. H. (1988). "Inadequacy of Property Specifications in ASTM C 270," Masonry: Materials, Design, Construction, & Maintenance, *ASTM STP 992*, edited by H. A. Harris, American Society for Testing and Materials, West Conshohocken, Pennsylvania, 7–17.

Skoulikidis, T., Tsakopoulos, A., and Moropoulos, T. (1986). "Accelerated Rebar Corrosion when Connected to Lightening Conductors and Protection of Rebars with Needles Diodes using Atmospheric Electricity," Corrosion Effects of Stray Currents and the Techniques for Evaluating Corrosion of Rebars in Concrete, *ASTM STP 906*, edited by V. Chaker, American Society for Testing and Materials, West Conshohocken, Pennsylvania, 15–291.

Thompson, N. G., and Beavers, J. A. (1990). "Effect of Ground-Water Composition on the Electrochemical Behavior of Carbon Steel: A Statistical Experimental Study," Corrosion in Natural Waters, *ASTM STP 1086*, edited by C. H. Baloun, American Society for Testing and Materials, West Conshohocken, Pennsylvania, 101–121.

Tilly, G. P., and Moss, D. S. (1982). "Long Endurance Fatigue of Steel Reinforcement," *IABSE Colloquium, Lausanne, Fatigue of Steel and Concrete Structures, IABSE Reports*, Vol. 37, International Association for Bridge and Structural Engineering, Zurich, Switzerland, 229–238.

Townsend, H. E., Jr. (1972). "Resistance of High Strength Structural Steel to Environmental Stress Corrosion Cracking," Stress Corrosion Cracking of Metals—A State of the Art, *ASTM STP 518*, American Society for Testing and Materials, West Conshohocken, Pennsylvania, 155–166.

Tuthill, L. H. (1966). "Hardened Concrete—Resistance to Chemical Attack," Significance of Tests and Properties of Concrete and Concrete Making Materials, *ASTM STP 169A*, American Society for Testing and Materials, West Conshohocken, Pennsylvania, 275–289.

Yi, F. L., and Carrasquillo, R. L. (1985). "A Study of the Thermal Behavior of Brick Under Service Conditions in a Structure," Masonry, Research, Application, and Problems, *ASTM STP 871*, edited by J. T. Conway and J. C. Grogan, American Society for Testing and Materials, West Conshohocken, Pennsylvania, 101–113.

5.0 REPORT OF STRUCTURAL CONDITION ASSESSMENT

The scope and content of the report should be consistent with the scope of the assignment. A simple checklist may suffice for a Cursory Assessment, and a letter report for a Preliminary Assessment. A full report would be appropriate for a Detailed Assessment.

This Section is a guide, since the order and content may vary with the scope of the engagement, the methods and techniques employed by the engineer, and for specific investigations. It provides a description of the various parts of the Preliminary and Detailed Assessments as may be used. The Section as written presents a form for a typical client/consultant relationship, but it does not preclude a different format. Appendix A is an example outline for such a report.

In the case of public agencies or multi-building owners where routine or special condition assessments are performed by staff personnel for the management of facilities, the form of the report may be established by that agency or owner. Similarly, such a client may require a specific format when the condition assessment is performed by a consulting engineer.

It must be recognized that any report could be used in future litigation. Hence, the degree of investigation and analysis should be stated explicitly. Conclusions and recommendations should be carefully worded, and the possible consequences of not following recommendations should be stated. A carefully worded disclaimer may be included in the report.

5.1 EXECUTIVE SUMMARY

An "Executive Summary" at the beginning of the report is discretionary. It would contain brief statements of the purpose, scope, conclusions, and recommendations.

5.2 INTRODUCTION

5.2.1 Purpose of Assessment
This introductory Section should be a concise statement describing the reasons for the condition assessment. Background information, if pertinent, may be related in this part and would include any applicable government/owner/user reporting requirements (see Section 1.2).

5.2.2 Scope of Investigation
The scope of the investigative work performed for the assessment will vary with the assignment and must be indicated specifically and clearly. Load assumptions and code jurisdictions should be indicated in this portion. Any unusual design features in life safety areas should be given special consideration.

5.2.2.1 Cursory Assessment
A cursory assessment may be used for multiple buildings to determine general condition and screening to establish priorities.

5.2.2.2 Preliminary Assessment
An initial "walk-through" visit for orientation and general impressions is common to most assessments. Review of available documents, further site inspections, preliminary analysis, and preliminary evaluation and recommendations may be part of this work (see Section 2.3). Results may indicate the need for an increase in the scope of the investigation.

5.2.2.3 Detailed Assessment
This is an expansion of the preliminary assessment, if needed. It would include a review of documentation, building inspection, materials assessment, detailed analysis, cost impact study, detailed evaluation, and recommendations (see Section 2.4).

5.2.2.4 Testing
The range and types of testing employed should be outlined in this Section.

5.2.3 Methods and Techniques
Methods and techniques employed in the survey, investigation, and testing should be covered in more detail in this part.

5.2.3.1 Data Collection and Documentation
Visual observations, photographs, oral and video tapes, measurements, drawings and sketches, and methods of investigation should be explained herein.

5.2.3.2 Testing
On-site and laboratory testing should be described.

5.2.4 Meetings
A summary of meetings during the investigative phase should be included.

5.3 DESCRIPTION OF STRUCTURE

5.3.1 General
A general description of the structure to be assessed should be given at this point. This description would include the type of architecture, type of structure, and materials comprising the structure.

5.3.2 Dates of Construction, Alteration, and Repair
Any information available should be given in this section.

5.3.3 History
The use or occupancy of the building throughout its life, maintenance procedures, environmental conditions, and any unusual behavior or other factors should be discussed.

5.3.4 Collected Data
Information accumulated during the survey, investigation, and assessment should be listed. This may include original drawings, insurance data, alterations, photographs and tapes, measured drawings, interviews, design calculations, etc.

5.4 DISCUSSION OF SITE VISIT

5.4.1 Overview
This is a report of the initial site visit for orientation and general impressions. On a limited assignment, this may comprise the entire engagement.

5.4.2 Survey
A more thorough survey should consider materials, real or inferred systems, dimensions, deflections and distortions, identification of problem areas, sample locations, and a record of data obtained.

5.4.3 Observations and Their Significance
This should be a summary of observations made, and how they affect the assessment.

5.5 PRELIMINARY OFFICE ANALYSIS

5.5.1 Computational Analysis
Describe the analytical methods used to utilize the collected information for the assessment of the building structure.

5.5.2 Code Conformance
Compliance with building codes, life safety requirements, and special owner/user criteria should be discussed for the initial construction, alterations, present condition, and potential future use (if applicable).

5.6 TEST PROGRAM

Testing methods employed shall be reviewed. Nondestructive, destructive, and load tests may be used.

5.7 FINAL COMPUTATIONAL ANALYSIS

This will be the basis for the evaluation of the structure.

5.8 INPUT FROM OTHER DISCIPLINES

The results of the survey and evaluation by other disciplines as they affect the building structure should be given here.

5.9 SUMMARY OF STUDY

Field and office work directed toward the condition assessment should be summarized. An "Executive Summary," including conclusions and recommendations, may be placed at the beginning of the report.

5.10 CONCLUSIONS AND RECOMMENDATIONS

Conclusions and recommendations are based upon the survey, investigation, testing, and evaluation. These require experience and "professional judgment." As such, they are not considered to be part of the Standard, although they are the most important part of the report. Prepare a recommended program if appropriate for maintaining the structural adequacy of the building.

5.11 APPENDICES

This should include all supporting data such as survey information, record drawings, photographs, test data and reports, computation, and references.

APPENDICES

APPENDIX A
REPORT OF STRUCTURAL
CONDITION ASSESSMENT
(Not part of Standard)

1. EXECUTIVE SUMMARY (OPTIONAL)
2. INTRODUCTION
    2.1 PURPOSE
        2.1.1 Determine current conditions/ deficiencies
            (a) Life safety
            (b) Performance
            (c) Serviceability
        2.1.2 Planning
            (a) Maintenance
            (b) Repairs
            (c) Budgeting
        2.1.3 Change of owner
        2.1.4 Change of occupancy
        2.1.5 Alterations or additions
        2.1.6 Code compliance
            (a) Life safety
            (b) Fire safety
            (c) Strengthening for lateral forces
        2.1.7 Adaptive reuse, rehabilitation, restoration
        2.1.8 Historic preservation
        2.1.9 Distress, failure or damage
            (a) System or material failure
            (b) Cracking, bulging
            (c) Water intrusion: rot, freeze-thaw damage, ice, corrosion
            (d) Fire
            (e) Flood
            (f) Wind: storm, tornado, hurricane
            (g) Blast, impact, vibration
            (h) Seismic event
            (i) Subsidence, differential settlement, sinkhole, heaving, swelling soils
            (j) Connection failure
            (k) Deterioration, erosion, weathering
            (l) Insect, rodent, or bird infestation
            (m) Others
    2.2 SCOPE OF INVESTIGATION AND ASSESSMENT
        2.2.1 Cursory assessment/screening
        2.2.2 Preliminary assessment
        2.2.3 Detailed assessment
        2.2.4 Survey
            (a) Corroboration of existing drawings
            (b) Measured drawings, measurement of structural members
            (c) Field evaluation of conditions (visual, probing, etc.)
            (d) Identification of problem areas
            (e) Record of observations (prints, matrix, photos, tapes)
        2.2.5 Testing
            (a) Nondestructive
            (b) Destructive
            (c) Load test
    2.3 METHODS AND TECHNIQUES
        2.3.1 Visual—including binoculars, magnifying glass, borescope, and fiber optics
        2.3.2 Photography, x-ray, infrared thermography, ultrasonic, impulse-echo, impulse-radar
        2.3.3 Tapes—oral, video
        2.3.4 Drawings and sketches
        2.3.5 Measurement
        2.3.6 Investigation procedure and tests used
    2.4 MEETINGS
3. DESCRIPTION OF STRUCTURE
    3.1 GENERAL
        3.1.1 Type of architecture
        3.1.2 Type of structure
        3.1.3 Materials
            (a) Masonry—stone, brick, CMU, clay tile, terra cotta, adobe, rammed earth
            (b) Wood—logs, hewn, sawn, laminated, treated, connections
            (c) Metals—iron, steel, aluminum, copper, bronze, lead
            (d) Concrete–placed, precast, plain, reinforced, prestressed
    3.2 History
        3.2.1 Dates of construction, alteration and repair
        3.2.2 Uses and occupancy—alterations, equipment, vibration, wear
        3.2.3 Environmental conditions—weather, heat, cold, chemicals, food products
        3.2.4 Site conditions
        3.2.5 Unusual loadings
    3.3 COLLECTED DATA
        3.3.1 Available drawings, specifications, calculations, reports
        3.3.2 Insurance descriptions, tax maps, deeds, building permits
        3.3.3 Alterations documentation
        3.3.4 Photographs and tapes

3.3.5 Measured drawings from survey
3.3.6 Interviews of people familiar with building
4. DISCUSSION OF SITE VISITS
   4.1 OVERVIEW—may be entire engagement
   4.2 SURVEY
      4.2.1 Materials
      4.2.2 Real or inferred systems
      4.2.3 Dimensions
      4.2.4 Deflections, distortions, deterioration
      4.2.5 Identification of problem areas
      4.2.6 Record of data
         (a) Sketches and drawings
         (b) Notes
         (c) Photographs, x-rays, infrared scans, etc.
         (d) Tapes—oral, video
   4.3 OBSERVATIONS AND THEIR SIGNIFICANCE
         (a) General
         (b) Limitations and qualifications
         (c) Need for immediate repairs
5. PRELIMINARY OFFICE ANALYSIS
   5.1 COMPUTATIONAL ANALYSIS
         (a) Determine live, dead, wind, seismic, and other appropriate loads
         (b) Structure geometry
         (c) Material properties; allowable working or ultimate strength parameters
         (d) Methods of analysis
         (e) Calculate load resistance and deformation
         (f) Determine need for material samples, member samples
   5.2 IDENTIFICATION OF TEST AREAS, APPROPRIATE TESTS
   5.3 CODE CONFORMANCE (PAST AND PRESENT)
6. TEST PROGRAM
   6.1 NONDESTRUCTIVE METHODS
   6.2 DESTRUCTIVE METHODS
   6.3 LOAD TEST
      6.3.1 System or component
      6.3.2 Methods
      6.3.3 Instrumentation
   6.4 RESULTS
7. FINAL ANALYSIS (Basis for evaluation)
8. INPUT FROM OTHER DISCIPLINES
9. SUMMARY OF STUDY
10. CONCLUSIONS
11. RECOMMENDATIONS

12. APPENDICES
    A. Survey information
    B. Record drawings
    C. Photographs
    D. Test data and reports
    E. Computations
    F. References

APPENDIX B
ORGANIZATION REFERENCES

American Association of State Highway and Transportation Officials (AASHTO)
American Concrete Institute (ACI)
American Institute of Steel Construction (AISC)
American Institute of Timber Construction (AITC)
American National Standards Institute (ANSI)
American Plywood Association (APA)
American Society for Metals (ASM)
American Society of Testing and Materials (ASTM)
American Society of Civil Engineers (ASCE)
American Welding Society (AWS)
American Wood Preservers Association (AWPA)
Brick Institute of America (BIA)
Building Officials and Code Administrators International (BOCA)
Construction Engineering Research Laboratory (CERL)
Federal Emergency Management Agency (FEMA)
Federal Highway Administration (FHWA), U.S. Department of Transportation
Forest Products Laboratory (FPL), U.S. Forest Service
International Association for Bridge and Structural Engineering (IABSE)
Institute of Civil Engineers
International Conference of Building Officials (ICBO)
International Union of Testing and Research Laboratories for Materials and Structures (RILEM)
Masonry Institute of America
Metals Engineering Institute
National Association of Corrosion Engineers
National Forest Products Association (NFPA)
National Institute of Standards and Technology (NIST), U.S. Department of Commerce
National Lumber Manufacturers Association

National Park Service, U.S. Department of Interior
National Particleboard Association (NPA)
National Science Foundation (NSF)
Naval Facilities Engineering Command (NFEC)
Northeast Lumber Manufacturers Association
Prestressed Concrete Institute (PCI)
Southern Building Code Congress International (SBCC)
Southern Forest Products Association
Southern Pine Inspection Bureau
Strategic Highway Research Program (SHRP)
The Masonry Society (TMS)
Transportation Research Board (TRB)
Truss Plate Institute (TPI)
U.S. Department of Agriculture
U.S. Department of Commerce (DoC)
U.S. Department of Housing and Urban Development (HUD)
U.S. Nuclear Regulatory Commission (NRC)
Welding Research Council
West Coast Lumber Inspection Bureau

INDEX

Absorption; concrete 8; masonry units 22, 71
Accident damage; concrete 94; structural steel 107; wood 125
Acidity; concrete 8
Acoustic emission; metal testing 19
Adhesion; masonry units 22
Adhesives; wood 31
Aggregate characteristics; concrete 94–95; mortar and grout 24
Air content; concrete 8, 38; mortar and grout 27
Air leakage; masonry 24
Alignment; mortar and grout 27
Alkali-aggregate reaction (AAR); concrete 94
Alkali-carbonate reaction; concrete 10
Alkali-silica reaction; concrete 10–11
Aluminum alloy 108
Anchorage; reinforcing steel 14, 16
Anchors; masonry 21, 26, 73
Angle to grain load; wood connectors 32
Animal attack; wood 32
Assessment 2; *See also* Structural materials assessment
Autoclave crazing; masonry units 22

Bed joint levelness; mortar and grout 27
Bed width variation; mortar and grout 27
Bend test; reinforcing steel 10, 52
Bending strength; wood 30
Bleeding channels; concrete 11
Blistering; masonry 24
Bowing; masonry 26
Bracing of compression elements and members; metals 17
Breaking strength; connectors 10, 17, 52
Buckling; metals 19
Buildings; defined 2; detailed assessment 6; evaluation 126, 133; load testing methods 87; load tests 126; overturning 133; sliding 133; stability 133; stiffness 126, 133
Bulging; masonry 26
Bulk specific gravity; masonry units 22, 71

Cable deterioration; causes 103–104
Calcite streak; masonry 24
Carbon content; metals 17; reinforcing steel 10
Carbonation; concrete 8, 11
Cast iron 108
Cement content; concrete 8, 39
Cement soundness 94
Cement-aggregate reaction; concrete 11

Chemical composition; connectors 17, 30; metals 17; reinforcing steel 10, 52
Chemical content; concrete 89
Chemical exposure; wood 31
Chemical resistance; masonry units 22
Chemical treatments; wood 31
Chipping; masonry 26
Chloride attack; concrete 11
Chloride content; concrete 9, 38
Client 2
Coating mass; metal connectors 17
Coating properties; reinforcing steel 10
Combined wood structural elements 29
Composite panels 28
Compression; wood 30
Compressive load; masonry assemblages 23
Compressive strength; concrete 9, 38; masonry units 22, 71
Computational analysis 2
Concentrated load; masonry assemblages 23, 71
Concrete; assessment 8–16; connections 10, 16, 104; connectors 10, 16, 104; deterioration 94–96; evaluation 94–104; function 94; physical conditions 10–16, 46–49, 99–100; properties 8–10, 38–40, 40–42, 97–98; reinforcing and tensioning steel 10, 14, 16, 96, 101–104; test methods 16, 38–40, 42–44, 46–49; *See also* Structural concrete assessment and evaluation
Condition; masonry flashing 26; metal connectors 20; lintels 26; mortar and grout 27; roofing 26; wood connector 32; wood glue 32
Connections; of concrete 10, 16, 104; defined 2; masonry 21; of metals 17, 20, 108; physical conditions 16, 20, 32; properties 10, 17, 30; for wood 30, 32, 126, 130–132
Connectors; of concrete 10, 16, 104, 111–114; masonry 21; of metals 17, 20, 108; physical condition 16, 20, 32–33, 113–114; properties 10, 17, 30–31, 111–112; for wood 30–31, 32–33, 126
Constituent portions; mortar and grout 24
Contaminated aggregate; concrete 12
Contaminated mixing water; concrete 12, 94–95
Control joints; masonry 26
Corrosion; cable steel 103–104; connectors 20, 32; masonry 24; metals 18; prestressed steel 102–103; prestressing steel 101; reinforcing steel 16, 96, 101; structural steel 104, 106, 107
Cost-impact study; detailed assessment 8; preliminary assessment 5
Cover; reinforcing steel 16

Coving; masonry 24
Cracking; concrete 11, 12, 13, 95; masonry 24, 26; metal connectors 20; metals 18; mortar and grout 27
Crazing; masonry 22, 24
Creep; concrete 9; wood 31
Critical systems; detailed assessment 7
Cross-sectional properties; concrete 12; connectors 20, 32; metals 18; reinforcing steel 16; wood 31
Crumbling; masonry 25
Cryptoflorescence; masonry 25

Damage from accidents; *See* Accident damage
Damp course; masonry 26–27
Dampproofing; masonry 27
Decay; wood 31, 80
Deflection; masonry 26; wood 31
Deformability; masonry assemblages 23, 71
Deformation; connectors 20, 32; metals 18; reinforcing steel 10, 52
Degradation; concrete 95
Delamination; concrete 12, 95; masonry 25; wood connectors 31
Density; concrete 9, 40; wood 29
Destructive testing 2
Detachment; masonry 26
Detailed assessment 6–8
Detailing; wood connectors 32
Deterioration; concrete 12; masonry 25
Dew point; masonry 25
Dimensional location; reinforcing steel 16
Dimensions; masonry units 22; metal connectors 20
Direct chemical attack; metals 18
Discoloration; concrete 12
Discontinuities; metal connectors 20
Disintegration; concrete 12
Displacement; masonry 26
Distortion; concrete 12; masonry 22, 26
Document review; detailed assessment 6; preliminary assessment 3
Drawings; detailed assessment 6
Driving rain index; masonry 25
Drying shrinkage; masonry units 22
Ductility; metals 17
Duration of load; wood 31–32, 125

Eccentricity; wood connectors 32–33
Efflorescence; concrete 12; masonry 22, 25, 72
Elasticity; concrete 9, 40; masonry 22, 23, 24; metals 17; wood 30, 80
Electrolytic or electrochemical corrosion; metals 18
Elongated bolt holes; wood connectors 33

Elongation; connectors 17, 30, 32; metals 17; reinforcing steel 10
Embrittlement; cable steel 103; prestressing steel 101–102
Erosion; concrete 12; masonry 25
Evaluation; concrete 94–104; defined 2; masonry 108, 115; metals 104–108; procedure 93–94, 126, 132; wood 115, 125–126, 127–132; *See also* Structural materials evaluation
Exfoliation; masonry 25
Expansion joints; masonry 27
Exposure; reinforcing steel 16
Exposure of joint reinforcement; mortar and grout 27

Fatigue; cable steel 103; concrete 95–96; prestressing steel 101; reinforcing steel 101; structural steel 107
Fatigue cracking; metals 18
Fatigue properties; metals 17
Fatigue strength; wood 30
Fire; reinforced concrete 96; wood 125
Fire protection; metals 18
Flashing; masonry 26
Flexural tensile strength; concrete 38; masonry assemblages 24, 72
Fracture cracking; metals 18
Fracture properties; metals 17
Freeze thaw damage; concrete 13, 14, 96
Freeze thaw properties; concrete 9, 40; masonry units 22, 72
Friability; masonry 25; mortar and grout 27

Geometry of structural components; metals 18
Geometry of structure; metals 18
Glulam 28
Grade; wood 29
Grout; deterioration 115; post-tensioned structures 102; properties 24, 115; *See also* Masonry and grout
Growth; wood 29

Hardness; connectors 17, 18, 30; mortar and grout 27
Heartwood 29
Honeycomb; concrete 13
Horizontal expansion joints; masonry 27
Hydrogen embrittlement; cable steel 103; prestressing steel 101–102

Imperviousness; masonry units 22
In situ stresses; metals 19
Initial rate of absorption (IRA); masonry units 22, 72
Insect attack; wood 32, 125

INDEX

Inspection 2
Interstitial condensation; masonry 25

Joints; masonry 26, 27
Juvenile wood 29

Knots; wood 29

Laminar tearing; metals 19
Laminated timber 28
Laminated veneer lumber (LVL) 28
Lateral resistance; wood connectors 30–31
Leaching; concrete 13
Length; metal connectors 20
Lintels; masonry 26
Load; buildings 87; masonry assemblages 23, 24
Load duration; wood 31–32, 125
Load tests 126
Loading and performance criteria; detailed assessment 7; preliminary assessment 3, 5
Location; masonry 27; metal connectors 20

Manufactured masonry units 21
Masonry; architectural masonry 26–27; connections 21; connectors 21; defined 20; deterioration 108, 115; manufactured masonry units 21; structural 26
Masonry assemblages; components 21, 108; physical condition 24–26, 76–78; properties 23–24, 118, 120–123
Masonry assessment and evaluation; anchors 21, 26, 73; assessment 20–27; evaluation 108, 115, 116–123; joint reinforcement 21; masonry assemblages 21, 23–26, 118; masonry deterioration 108, 115; masonry units 21, 22–23, 116–117; mortar and grout 21, 24, 115, 119, 124; natural building stones 21; physical conditions 24–27, 76–78, 120–123; properties 22–24, 71–73, 116–118; reinforcing steel 21; test methods 27, 71–73, 76–78; ties 21, 26
Mass determination; masonry units 22
Material properties; detailed assessment 7; evaluation 126; preliminary assessment 5
Member analysis; detailed assessment 7; preliminary assessment 5
Metals; *See* Aluminum alloy; Cast iron; Metals assessment and evaluation; Steel; Wrought iron
Metals assessment and evaluation; assessment 16–20; connections 17, 19, 108; connectors 17, 19, 111–114; evaluation 104–108, 109–114; physical conditions 17–19, 56–58; properties 17, 52–54; test methods 20, 52–54, 56–58, 60–63, 65–69

Modulus of elasticity; concrete 9, 40; masonry 22, 23, 24; metals 17; wood 30
Modulus of rupture; concrete 9; masonry 22, 24, 72
Moisture content; concrete 9; masonry units 22–23; wood 32, 80, 125
Moisture effects; wood 32
Moisture expansion; masonry units 23
Mortar and grout 21; deterioration 115; physical conditions 124; properties 24, 119

Natural building stone 21
Net area; masonry units 23
Nonbinding condition; metal connectors 20
Nondestructive testing (NDT) 35; of concrete 42–44, 46–49; defined 2; of masonry 76–78; of metals 56–58; of steel 65–69; of wood 82–84
Nonstructural components; defined 2; detailed assessment 7; preliminary assessment 5

On-site load testing 2
Opacity; masonry units 23
Overall or local buckling; metals 19
Overstressing; metals 19
Overturning; evaluation 133

Paste characteristics; mortar and grout 24
Peeling; masonry 25
Penetration of nails and screws; wood connectors 33
Permeability; concrete 9
Permeance; masonry 25–26
Physical damage; wood 32
Pitting; masonry 25; mortar and grout 27
Plumbness; masonry 26
Plywood 28
Popouts; concrete 13, 15
Porosity; metal connectors 20
Post-tensioning steel; physical conditions 14, 16, 56–58, 106; properties 10, 52–54, 105
Pre-tensioning steel; physical conditions 14, 16, 56–58, 106; properties 10, 52–54, 105
Preservatives; wood 80
Prestressing steel; deterioration 101–103
Professional engineer 2
Profile; metal connectors 20
Proof load; connectors 17, 31
Proportions of aggregate; concrete 9
Pullout strength; concrete 9

Reduction of area; connectors 17; metals 17; reinforcing steel 10
Reinforcing steel; deterioration 96, 101; physical conditions 14, 16, 56–58; properties 10, 52–54

145

Report 5–6, 7–8, 137–139; description of structure 138; detailed findings 7; executive summary 138, 140; introduction 138; sections 138–141
Resistance to delamination; wood connectors 31
Resistance to freezing and thawing; concrete 9, 40
Rising damp; masonry 25
Roofing; masonry 26

Sagging; masonry 26
Salt fretting; masonry 25
Sampling; evaluation 126; masonry units 23
Sandiness; mortar and grout 27
Sapwood 29
Saturation coefficient; masonry units 23, 73
Scaling; concrete 13–14, 15, 95
Secant modulus of elasticity; masonry 23, 24, 72
Settlement; masonry 26
Shape; reinforcing steel 16
Shear strength; masonry assemblages 24, 73; wood 30
Shelf angles; masonry 26
Site inspection; detailed assessment 6; preliminary assessment 3
Size; masonry units 23; metal connectors 20
Slag deposit; metal connectors 20
Sliding; evaluation 133
Slotted holes for movement; metal connectors 20
Smoothness; metal connectors 20
Solid sawn wood 28
Solubility; masonry units 23
Soundness; concrete 9; metal connectors 20
Spalling; concrete 14, 95; masonry 25, 26
Specific gravity; masonry units 22, 71; wood 81
Splits at connection; wood connectors 33
Splitting tensile strength; concrete 9, 40; masonry units 23, 73
Stability; evaluation 133
Staining; masonry 25
Steel; cable deterioration 103–104; physical conditions 66–68; properties 65; properties on metal specimens 60–63; structural steel 104–106; *See also* Post-tensioning steel; Pre-tensioning steel; Reinforcing steel
Stiffness; evaluation 126, 133
Stratification; concrete 14
Strength of connections; reinforcing steel 10, 53
Stress at extension; metal connectors 17
Stress corrosion; cable steel 103–104; prestressed steel 102–103; structural steel 107
Structural components; defined 2; preliminary assessment 5
Structural composite lumber 28

Structural concrete assessment and evaluation; assessment 8–16; concrete physical conditions 10–16, 46–49, 99–100; concrete properties 8–10, 38–40, 40–42, 97–98; connections 10, 16; connectors 10, 16; deterioration causes 94–96; evaluation 94–104; reinforcing and tensioning steel 10, 96, 101–106; test methods 16, 38–40, 42–44, 46–49
Structural condition assessment; agreements 1–2; detailed assessment 6–8; detailed findings 7; flowchart of activities 4; preliminary assessment 3, 5–6; procedure 3–8; purpose 1; report 5–6, 7–8, 137–139; *See also* Structural materials assessment; Structural materials evaluation
Structural evaluation; detailed assessment 7; preliminary assessment 5
Structural glued laminated timber (glulam) 28
Structural materials assessment 8–36; concrete 8–16, 38–40, 42–44, 46–49; masonry 20–27, 71–73, 76–78; metals 16–20, 52–54, 56–58, 60–63, 65–69; procedure 3–8; test methods 35–87; wood 27–33, 80–84
Structural materials evaluation; concrete 94–104; masonry 108, 115; metals 104–108; procedure 93–94, 126, 132; wood 115, 124–126, 127–132
Structural materials testing; *See* Test methods
Structural performance; concrete 14
Structural steel; deterioration 104, 106; function 104; physical conditions 110; properties 109
Structural system; detailed assessment 6
Structural wood panel products 28–29
Sugaring; masonry 25
Sulfate attack; concrete 14, 96
Surface condensation; masonry 25
Surface crust; masonry 25
Surface induration; masonry 25
Surface temperature; masonry 25
Surface texture; masonry units 23

Tensile load; masonry assemblages 24, 73
Tensile strength; concrete 9–10, 38, 40; connectors 17, 31; masonry assemblages 24, 72, 73; metals 17; reinforcing steel 10; wood 30
Tensile strength of filler material; metal connectors 17
Test methods 35–36; concrete assessment 16, 38–40, 42–44, 46–49; evaluation 93–94; load tests 87, 126; masonry assessment 27, 71–73, 76–78; metals assessment 20, 52–54, 56–58, 60–63, 65–69; on-site load testing 2; wood assessment 33, 80–84; *See also* Destructive testing; Nondestructive testing

Testing laboratory 2
Thermal expansion; masonry units 23
Tidemark; masonry 25
Ties; masonry 21, 26
Tightness; connectors 20, 33
Timber 28
Timber joint compressive strength 31
Timber joint tensile test 31
Timber trusses 28
Transition area; masonry 25
Transverse load; masonry assemblages 24, 73
Trusses; wood 28

Ultrasonic probes; metal testing 19
Uniformity; metal connectors 20
Uniformity of concrete 14
Uniformity of mix; concrete 10
Unsound cement 14
Unsound concrete 14

Vertical load capacity test; wood 31
Void area in cored units; masonry units 23
Voids; mortar and grout 27

Warpage; masonry 23, 25, 26
Water absorption; masonry units 23
Water penetration/permeance; masonry 25–26, 73
Water-cement ratio; concrete 10
Weathering; masonry 26; wood 32
Weld shear strength; reinforcing steel 10, 53
Withdrawal strength; wood connectors 31, 80
Wood 27; deterioration 115, 125
Wood assessment and evaluation; assessment 27–33; connections 30, 32, 126; connectors 30–31, 32–33, 126, 130–132; evaluation 115, 125–126; manufactured wood products 28–29; physical condition 31–32, 82–84, 126, 128–130; properties 29–30, 80–81, 126, 127; test methods 33, 80–84; wood deterioration 115, 125
Wood and wood products 27, 28–29
Wrought iron 108

Yield strength; metals 17; reinforcing steel 10; wood connectors 31
Yield strength of filler material; metal connectors 17
Young's modulus; masonry units 23